MATHEMATICIANS OF THE WORLD, UNITE!

THE INTERNATIONAL CONGRESSES OF MATHEMATICIANS

ZURICH, AUGUST 9–11, 1897

PARIS, AUGUST 6–12, 1900

HEIDELBERG, AUGUST 8–13, 1904

ROME, APRIL 6–11, 1908

CAMBRIDGE, AUGUST 22–28, 1912

STRASBOURG, SEPTEMBER 22–30, 1920

TORONTO, AUGUST 11–16, 1924

BOLOGNA, SEPTEMBER 3–10, 1928

ZURICH, SEPTEMBER 5–12, 1932

OSLO, JULY 14–18, 1936

CAMBRIDGE (MA), AUGUST 30–SEPTEMBER 6, 1950

AMSTERDAM, AUGUST 2–9, 1954

EDINBURGH, AUGUST 14–21, 1958

STOCKHOLM, AUGUST 15–22, 1962

MOSCOW, AUGUST 16–26, 1966

NICE, SEPTEMBER 1–10, 1970

VANCOUVER, AUGUST 21–29, 1974

HELSINKI, AUGUST 15–23, 1978

WARSAW, AUGUST 16–24, 1983

BERKELEY, AUGUST 3–11, 1986

KYOTO, AUGUST 21–29, 1990

ZURICH, AUGUST 3–11, 1994

BERLIN, AUGUST 18–27, 1998

BEIJING, AUGUST 20–28, 2002

MADRID, AUGUST 22–30, 2006

MATHEMATICIANS OF THE WORLD, UNITE!

The International Congress of Mathematicians
A Human Endeavor

Guillermo P. Curbera

CRC Press
Taylor & Francis Group
Boca Raton London New York

CRC Press is an imprint of the
Taylor & Francis Group, an **informa** business

AN A K PETERS BOOK

CRC Press
Taylor & Francis Group
6000 Broken Sound Parkway NW, Suite 300
Boca Raton, FL 33487-2742

First issued in paperback 2019

© 2009 by Taylor & Francis Group, LLC
CRC Press is an imprint of Taylor & Francis Group, an Informa business

No claim to original U.S. Government works

ISBN-13: 978-1-56881-330-1 (hbk)
ISBN-13: 978-0-367-38596-5 (pbk)

Library of Congress Cataloging-in-Publication Data

Curbera, Guillermo P.
    Mathematicians of the world, unite! : the International Congress of Mathematicians : a human endeavor / Guillermo P. Curbera.
       p. cm.
    Includes bibliographical references and index.
    ISBN 978-1-56881-330-1 (alk. paper)
    1. International Congress of Mathematicians–History. 2. Mathematics–Congresses–History. I. Title.

QA1.C853 2008
510–dc22

2008006190

Visit the Taylor & Francis Web site at
http://www.taylorandfrancis.com

and the CRC Press Web site at
http://www.crcpress.com

CONTENTS

FOREWORD ix

PREFACE xi

INTRODUCTION xv

THE ORIGINS 1

I   EARLY TIMES 9

ZURICH 1897 11

PARIS 1900 19

HEIDELBERG 1904 27

ROME 1908 37

CAMBRIDGE 1912 47

   Interlude—IMAGES OF THE ICM 55

II   CRISIS IN THE INTERWAR PERIOD 67

STRASBOURG 1920 69

TORONTO 1924 75

BOLOGNA 1928 83

ZURICH 1932   91

OSLO 1936   101

   Interlude—AWARDS OF THE ICM   109

III   THE GOLDEN ERA   125

CAMBRIDGE, MA, 1950   127

AMSTERDAM 1954   133

EDINBURGH 1958   139

STOCKHOLM 1962   143

A PERIOD OF SUCCESS   149

   Interlude—BUILDINGS OF THE ICM   155

IV   ON THE ROAD   173

MOSCOW 1966   175

NICE 1970   185

VANCOUVER 1974   191

HELSINKI 1978   199

WARSAW 1982   207

BERKELEY 1986   219

   Interlude—SOCIAL LIFE AT THE ICM   227

V   IN A GLOBAL WORLD   255

KYOTO 1990   257

ZURICH 1994   267

BERLIN 1998   275

BEIJING 2002   285

MADRID 2006   293

    Coda—INTERNATIONAL MATHEMATICAL UNION   305

ACKNOWLEDGMENTS   307

BIBLIOGRAPHY   309

INDEX   315

This book has been published with the support of the University of Sevilla.

FOREWORD

MY FIRST ICM was the congress at Harvard 1950. I was then 22 years old and had just defended my thesis in classical complex analysis at the University of Uppsala in Sweden. This was to be my first contact with non-Swedish mathematicians. The voyage was a major enterprise—it took ten days.

The congress was a great experience. I was amazed to encounter the richness of our field and how unimportant my own specialty was considered by many people. I made friends from different parts of the world, and these contacts have lasted through the years. I saw icons of mathematics whose names I knew from theorems and listened to their lectures. I remember in particular Jacques Hadamard. The organizers of the congress had with great effort managed to get a visa for him for a few days, in spite of the risk for the security of the country to let an 85-year-old communist in. This was my first contact with the problem of how politics interferes with mathematics. Much more on this subject can be found in this book. At the congress, I listened to lectures by not only Hadamard but also H. Cartan, K. Gödel, J. Leray, J. von Neumann, and S.-S. Chern, to just mention a few. I also remember the excitement of the Fields Medals—who would win?—and the discussions afterwards.

Much is the same today, and much has changed. For us who are now old, it is a great trip of nostalgia to remember the congresses and see the faces of the people on the following pages.

We must be grateful to the author for collecting so many pictures from the past 100 years. The book complements the earlier book by D. Albers, G. Alexanderson, and C. Reid (1987). But much has changed. Mathematicians now meet on a regular basis, travel is easy, and information is instantaneous over the Internet. The specialization within mathematics is very pronounced. The meetings are specialized as well as the journals, and many people are as mathematically isolated as I was in Sweden during and after the war. The ICMs, however, still provide the unique possibility for the young to see today's icons, to learn and respect areas of the field other than their own specialty. We often talk about the unity of mathematics, and the ICMs give us the possibility to get new impulses from areas that we otherwise don't see.

I welcome this book and hope it will inspire international cooperation without politics and promote curiosity and interest in the whole area of mathematics.

Lennart Carleson

PREFACE

THE ORIGIN OF THIS BOOK lies in the celebration of the International Congress of Mathematicians in Madrid in 2006. The congress was the 25th in the series of international congresses. The Executive Committee of ICM 2006 decided to celebrate the occasion with an exhibit on the history of the congresses, and I was invited to organize it. The idea was to make a largely graphic display, showing the 25 official posters and some photographs of mathematicians. It seemed nice and simple.

The main source of inspiration for the exhibition was the two books on the subject (always books!). One was the original *International Mathematical Congresses: An Illustrated History 1893–1986*, by Donald J. Albers, Gerald L. Alexanderson, and Constance Reid. The other book was the precise and encyclopedic *Mathematics without Borders: A History of the International Mathematical Union*, by Olli Lehto. I reread both books and then contacted Lehto in order to visit the archives of the International Mathematical Union. Lehto had organized the archives at the University of Helsinki after leaving his post as Secretary of the Union. This was in April 2004.

Naively enough, I went to Helsinki with the idea of coming back with the exhibition in my bag. I spent a week there, hosted by Lehto, who granted me full access to the Union's archives. Then came the surprise: there was almost no usable material for the planned exhibition. There were hundreds, thousands of memoranda, letters, and invoices but no trace of any poster or photograph. The exception was the 1978 International Congress that Lehto had organized in Helsinki. Even worse, there were no records at all of the congresses before World War II.

Back to the beginning again. I realized that there was only one solution: reading the proceedings of all the congresses in order to find the thread that would weave together the script of the exhibition. The experience was fascinating. The 24 congresses have produced 54 volumes, the bulk of which are devoted to the invited lectures and communications, but there was a great deal of very interesting information about the course of the congresses. I had found the thread, so I wrote the script for the exhibition.

Next came the problem of finding the images that would support the script. This is where the mathematical community as a whole came into action. I will illustrate how I acquired the graphic materials with the story of three sets of photographs.

Let us start with the 1962 International Congress held in Stockholm. I was obsessed with the idea of showing a photograph of the King of Sweden presenting the Fields Medals to Lars Hörmander and John Milnor. I contacted Hörmander and the organizers of the congress, Lars Gårding and Lennart Carleson, but they did not have any photographs.

Carleson was very interested in my request and directed me to other scientific institutions in Sweden. I even contacted the photographic agency that deals with matters of the Swedish Royal House. Nothing at all: I was defeated. Some months later, I received a message from Anne Miche de Malleray, archivist of the Center for History of Science of the Royal Swedish Academy of Sciences, one of the many institutions that I had contacted during my search. The message said: "Archival work sometimes moves in mysterious ways. Today when I was looking into a completely different matter I came across an unlabeled box. When I opened it I saw six pictures from the 1962 ICM, and among them two showing the presentation of the Fields Medals by King Gustaf VI Adolf." Finally, here was success.

Another obsession of mine was the 1966 International Congress, held in Moscow. I was especially interested in finding photographs of this congress, since they would show the world of the communist Soviet Union that no longer exists. I contacted the organiz-

ers of the congress; again I had no success. When I attended a conference in Moscow, I visited the Steklov Institute looking for any material from the congress; I was not lucky. In desperation, I asked for help from Anna Doubova, a Russian colleague at my university. We searched endlessly through the Web in the archives of Russian scientific institutions with no success. About to give up, Anna suggested searching wildly through the Internet. My opinion was that for this type of information, the Internet was useless, and I was right... until Anna suggested searching with Cyrillic characters. After a while, we found more than we had expected. On the personal Web page of Serge I. Khashin, from the Faculty of Mathematics of the University of Ivanovo, there were dozens of photographs from ICM 1966. The department had a rather large archive with negatives belonging to the late S. V. Smirnov, who had attended the 1966 congress and had taken his own personal pictures. We had found a treasure!

1932 congress card of J. J. Burckhardt. (Courtesy of J. J. Burckhardt.)

Sehr geehrter Herr
Cimbora.

Besten Dank für den Brief von
17 Mai:
Beiliegend sende ich alles
was ich von Congress 1932
besitze.
Nicht retournieren.
Meine Hilfe ist weg, muss alles
selbst erledigen!!
beste Grüsse
J-J. Burckhardt
(geb. 1863).

Letter by J. J. Burckhardt. (From the author's personal files.)

The last story concerns the International Congresses held in 1932 and 1994 in Zurich. In the proceedings of ICM 1994, I saw the photograph of J. J. Burckhardt, who had helped to organize the 1932 congress and also had attended the 1994 congress. I contacted a friend, Hans Jarchow from the University of Zurich, who had been on the organizing committee of the 1994 congress, and inquired about Burckhardt. He was living up in the Alps, but he was very old, and I was told that he should not be disturbed. Nevertheless, I contacted Burckhardt in February 2006, and after an exchange of letters, one day I received a large envelope with a handwritten note in which Burckhardt explained that he was sending me his archives from the 1932 congress. The envelope contained photographs, newspaper clippings, and documents from ICM 1932 (see, for example, page xii). In the letter, under his signature, Burckhardt had written "Geboren 1903" (born 1903). In December 2006, the sad news came from Zurich that Burckhardt had passed away.

These and many other stories illustrate the way in which the exhibition became a collective effort. The exhibition was entitled *The ICM through History*. The construction was a navy blue cube, ten by eight meters at its base and three meters high (see page xiv). It contained nearly 500 images of different types, as well as many stories. A more detailed description of the exhibition can be found in Part V of this book, "In a Global World," where the Madrid 2006 International Congress is discussed.

This book gives the full description of what was prepared for the exhibition. Sir John Ball from Oxford, who was president of the International Mathematical Union when the 2006 International Congress took place, said that "[t]he tradition of the ICM is a wonderful one." This book intends to substantiate that statement. As a follow-up to the exhibition, the book can be considered, in some sense, an offspring of a collective effort.

It is a pleasure to acknowledge the support that I have received. The greatest support came from members of the Executive Committee of ICM 2006: President Manuel de León, Emilio Bujalance, Antonio Durán, and Rosa Echevarría. Antonio Durán deserves special notice as a friend who believed in this project from the beginning. Olli Lehto has always helped and advised me. Juan Arias de Reyna has shared with me his wide knowledge. Two colleagues from my department have helped me beyond all expectations: Pedro López and Olvido Delgado. I am indebted to Eileen O'Brien, who counseled me beyond the call of friendship. José Luis Arántegui has helped me to navigate through the Babel of languages of the proceedings of the congresses. And last but not least, I thank my wife Lourdes, not only for her constant support, but also for the extra support at critical moments. In March 2006,

The exhibition *The ICM through History* at the Madrid 2006 International Congress of Mathematicians. (From the author's personal files.)

the ICM Committee received unfortunate news: financial support from the Spanish government was much less than had been expected. The president of the ICM committee called everybody with one message: cut down expenses and freeze projects. The first in line for this was the exhibition. After two years of work, there might be no exhibition. At that moment, I made a difficult decision: I would keep working, no matter what. If the exhibition was to be buried, it would be so completely finished. These difficulties were overcome, and the exhibition was opened at the same time as the International Congress, in August 2006.

I am proud to recognize that the list of people and institutions to whom I owe gratitude is much longer; the complete list is printed in the final pages of the book.

INTRODUCTION

The series of International Congresses are very loosely held together. They are not congresses of mathematics, that highly organized body of knowledge, but of mathematicians, those rather chaotic individuals who create and conserve it.

R EADING THE ABOVE TEXT produces different reactions in the reader, because not many mathematicians recognize themselves as "rather chaotic individuals," although they do consider themselves among those who "create and conserve" mathematics. The person who made this observation was not a romantic poet nor an outsider to mathematics, but rather the U.S. mathematician Oswald Veblen, who was president of the American Mathematical Society between 1923 and 1924 and presided over the first International Congress of Mathematicians held after World War II in 1950 at Harvard University. The occasion where he expressed this view of the ICMs and their significance was a solemn one: September 2, 1954, in Amsterdam's Concertgebouw (concert hall) at the opening of the International Congress. Following a well-established tradition, in the opening ceremony the president of the preceding congress proposed to the participants the election of the new president of the congress. The nominee was usually the person who had presided over the organizing committee. In the meantime, he was the remaining representative until the next congress. At that time, the International Mathematical Union had

not yet taken over the control of the ICM and, as Veblen explained,

> At each congress they somehow agree on the country where the next one is to be held and then leave it to their colleagues in this country to work out a program.

The importance of the human factor in the ICMs, noted by Veblen and which can be appreciated in the peculiar mechanism of the succession of presidents, has had many consequences for the history and the character of the ICMs. One of them is that the sequence of the congresses reflects very accurately the world in which they have taken place, from political events to economic development to social habits. This situation has given the international congresses a dual personality. On one hand, they are scientific summits of mathematics where every four years a steady image of the state-of-the-art is displayed, with its achievements and its challenges. This facet of the ICMs is the one we see in the scientific content of the congress proceedings. Invited lectures, laudations for the awardees, and communications draw a lively and kaleidoscopic mosaic of the mathematics of the moment.

But it is the human factor that has given life and vitality to another facet, to which we will refer as the cultural side of the ICMs. This human facet and its journey, mostly through the twentieth century, is the aim and focus of this book, and it explains the way in which we look at the congresses.

We enjoy reading the original words spoken in the congresses; they allow us to grasp the viewpoints of the time and the world in which they were said. Sometimes they are so vivid that we can even imagine we are listening to them. We believe what is said in the congress proceedings. That is why when someone wrote, "A beautiful and inspired song was played," we write that down. These are the reasons for the extensive quotes; we think that the reader will enjoy reading them.

The same happens with countries, cities, places, cultures, and languages. Through the narration of the congresses we see the world evolve; we detect the glories and obsessions of different countries. Each language carries an interpretation of the world; discovering these interpretations is an adventure. Titles and pompous expressions illuminate, or shade, a truth. Names of places, hotels, institutions, and songs have stories attached to them and represent national character.

The relevance of the ICMs as social events in the countries and cities where the congresses have been held can be gauged by the importance of the buildings that have hosted the congresses. We look at these buildings; they display a remarkable review of the history of architectural styles.

The lists of plenary lectures can be somewhat long and even tedious, but they are interesting in that they help us sense the flow of time in mathematics. In some, we find topics of interest today; in others, the hot topics of the time are now forgotten or neglected. As time goes by, the titles of the lectures lose their charm and become more technical, sometimes even cryptic. A similar phenomenon happens with the names of the lecturers and participants. We are moved when we recognize well-known mathematicians, and at the same time surprised and puzzled not to recognize others. The same occurs with the list of scientific sections into which each congress divided mathematics. They reveal a view of mathematics—the strength of certain fields and the decline of others—throughout the years.

The picture drawn is not uniform; it never could be. Some congresses were much more intense than others. Some left a detailed record of their course; others did not.

The role of pictures in the book is crucial. They are not just a "side dish." With them, we have the same aim as with the original texts: to imagine the world in which they were taken. Unless stated, the pictures are authentic; that is, they are from the place and time described. They should not be treated as fast food that one gobbles without tasting. Rather, they should be used as a vehicle for imagining what they do not show.

What is the story of the ICMs that this book tells? The published proceedings of the Amsterdam 1954 congress have a section entitled "A Bird's-Eye View of the Congress." This is our intention: to give a bird's-eye view of the congresses. This bird's-eye view consists of two intertwined stories: a chronological narration and a thematic story of the congresses.

We follow the historical course of the congresses in chronological order. The 25 congresses fall naturally into five periods, determined by important historical events. Each period included congresses with a unifying character. "The Origins" briefly surveys the scientific, social, and historical conditions that made possible—and necessary—the appearance of the international congresses. "Early Times" looks at the period of progressive consolidation of the congresses in the years 1897–1912. In "Crisis in the Interwar Period," we follow the struggle of the congresses during the years 1920–1936, confronting the aftermath of the Great War. In this period, nonscientific influences were very strong in the international congresses. "The Golden Era" is devoted to the years 1950–1962, when the splendor of the congresses was deployed in full and when the classic congresses took place. "On the Road" narrates the congresses in the period 1966–1986, when attendance became popularized. The last period, 1990–2006, is presented in "In a Global World," where we find new lines of development emerging in the ICMs.

Intertwined with the chronology are the interludes. These give a somewhat different story of the ICMs, not following the flow of time but focusing on a particular feature of the congresses. Their role is that of a celery sorbet in a degustation menu: cleansing taste and preparing the palate for the next course. With this aim, "Images of the ICM" looks at the graphic creations used by the congresses, appearing in posters, logotypes, stamps, and other printed materials. "Awards of the ICM" is devoted to the awards associated with the international congresses: the Fields Medal, the Nevanlinna Prize, and the Gauss Prize. "Buildings of the ICM" is a visual tour through some of the buildings around the world where the ICMs have been hosted. The most important of these interludes (a true main course) is "Social Life at the ICM," where we review how the mandate set in the first congress of "fostering personal relations between mathematicians of different countries" has been accomplished.

We end with a coda devoted to summarizing the turbulent life of the International Mathematical Union.

THE ORIGINS

It is true that most of the great ideas of our science have raised and matured in the silence of the working studio; no other science, but possibly for Philosophy, presents a character so eremitic and secluded as mathematics. And yet, in the heart of a mathematician lives the necessity for communicating and expressing himself to his colleagues. And each of us certainly knows by personal experience how stimulating personal scientific intercourse can be.

THUS ADOLF HURWITZ addressed the participants in the first International Congress held in Zurich in 1897. The point raised by Hurwitz, although obvious and well-known to mathematicians, might surprise the general public or even other scientists. Mathematics has traditionally been pictured as an abstruse, austere, and solitary science. This image arises from the combination of its encrypted expression together with a meek acceptance of its veracity by the layman. Also contributing to this image is a gallery of pictures of mathematicians ranging from the absent-minded savant to the daft scientist. This is how Livy showed us Archimedes, "leaning over some drawing he had made on the ground" as he is being murdered by the soldiers of the Consul Marcellus in Syracuse (see page 2). Or we may remember Georg Cantor continually visiting the *Nervenklinik* at the end of his life, or Kurt Gödel starving himself to death at Princeton.

The tour we are about to begin through the series of the International Congress of Mathematicians will reveal that deep in the heart of mathematics there is a strong impulse towards communication and an intense sentiment of building a community. In a sense, the international congresses are the highest example of this community identity. This vision stands in stark contrast to the general image of this science.

How did these international congresses come to be? Why did they arise at the end of the nineteenth century, and not earlier or later? The international congresses were the last step of a long process. A bit of history is needed to follow the steps of this process.

Up to the eighteenth century, practitioners of science—of mathematics, in particular—did not enjoy a good professional base. For them, science was a passionate but secondary dedication. Under the patronage of the powerful modern states, by the end of the seventeenth century, the first scientific academies were created. In the 1660s, the Royal Society of London and the Académie Royale des Sciences in Paris were founded. Near the beginning of the eighteenth century, the Societas Regia Scientiarum (known as the Berlin Academy), in 1700, and the Academia Scientiarum Imperialis (known as the St. Petersburg Academy), in 1724, came into being. These institutions hosted—in marvelous solitude—the eighteenth-century mathematicians Leonhard Euler, Daniel and Nicolas Bernoulli, Jean d'Alembert, Joseph Louis Lagrange, and others.

The death of Archimedes. (Courtesy of the Städtische Galerie Liebieghaus, Frankfurt am Main.)

Then, the French Revolution woke Europe from its lethargy. Some of its many consequences were the profound changes in education and science (in France, all universities were closed down, and the École Polytechnique was created). From the reactions throughout Europe to the Revolution, a new system of higher education and scientific research arose from the model outlined by Wilhelm von Humboldt, for the newly founded University of Berlin and other Prussian universities. Thus, science moved to the universities; its development joined with higher-level education. University professors assumed the additional (and imperative) role of researchers. This new situation caused the establishment of new academic chairs devoted entirely to mathematics, and of new types of academic positions. Important innovations were the mathematical seminars in the universities and the reference libraries with specialized literature, where students were trained intensively. Job opportunities multiplied, and in some years the number of professional mathematicians also increased.

Another important feature was scientific journals. The first journals appeared during the Scientific Revolution in the second half of the seventeenth century; these were published mainly by academies and learned societies. They were the *Journal des Savants*, privately founded in France in 1665, and the *Philosophical Transactions of the Royal Society*, founded in London also in 1665. Next were the *Giornale de' letterati*, Rome, 1668, and *Acta Eruditorum*, Leipzig, 1682, where Leibniz published many of his papers on the newly created subject of differential calculus. During the eighteenth century, numerous other academic journals were founded, in which mathematical articles intermingled with articles in other disciplines. The increasing specialization of science towards the end of the century led to the rise of subject-oriented journals. At first, pure mathematics appeared together with its fields of application—physics, astronomy, and geography. Later, the first third of the nineteenth century witnessed the creation of

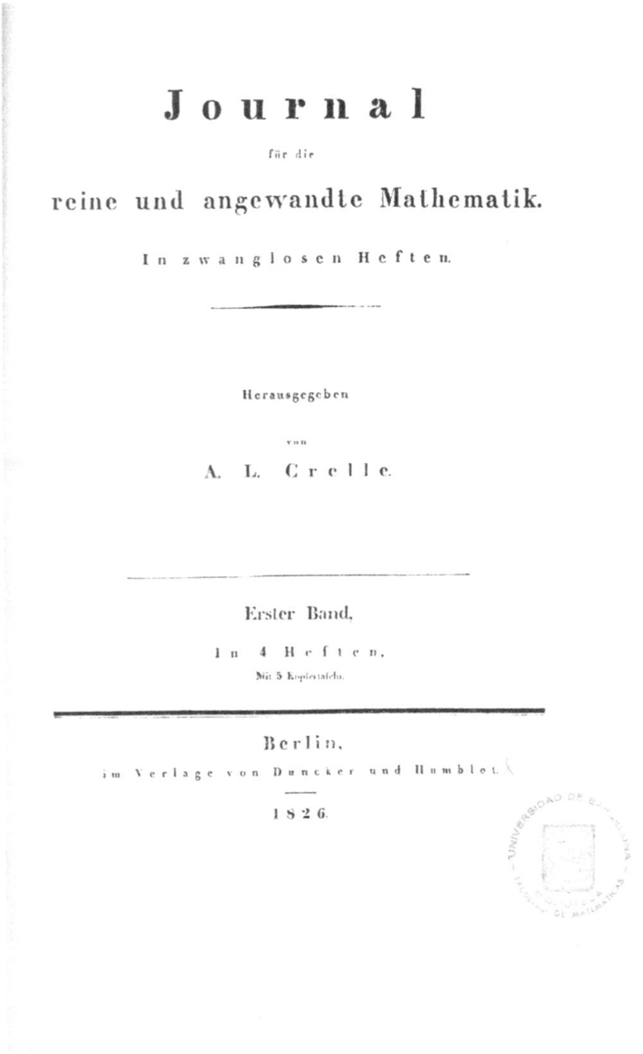

Journal

für die

reine und angewandte Mathematik.

In zwanglosen Heften.

Herausgegeben

von

A. L. Crelle.

Erster Band,

In 4 Heften.

Mit 5 Kupfertafeln.

Berlin,

im Verlage von Duncker und Humblot.

1826.

Crelle's Journal. (Courtesy of the Biblioteca de la Universidad de Barcelona.)

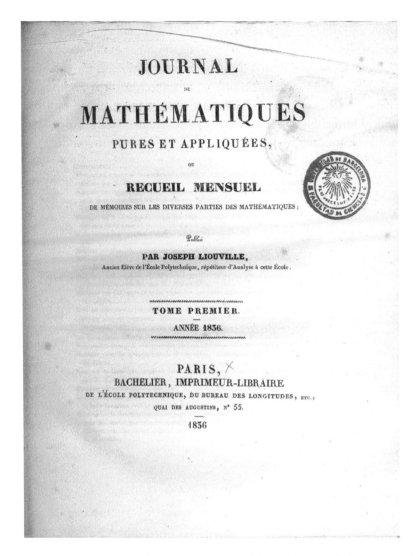

The *Journal de Mathématiques Pures et Appliquées*. (Courtesy of the Biblioteca de la Universidad de Barcelona.)

research journals devoted solely to mathematics. The *Annales de Mathématiques Pures et Appliquées*, founded in 1810 by Joseph Diaz Gergonne, was first, only published until 1832. Next was the *Journal für die reine und angewandte Mathematik*, founded by August Leopold Crelle (engineer, mathematician, and consultant to the Prussian Ministry of Instruction) in 1826. It still exists today as a prestigious journal, known as *Crelle's Journal* (see page 3). A third journal, which also exists today, was the *Journal de Mathématiques Pures et Appliquées*, founded by Joseph Liouville in 1836.

Another step was the creation of national mathematical societies in the second half of the century. There was an imperative for many countries to create such new institutions. Their concerns had not been covered by the academies, which were too elitist and had a limited number of members. These new necessities were the growing number of professional mathematicians, their increasing productivity—which had caused the fragmentation of mathematics—and the unsatisfactory possibilities for scientific communication. Such societies already existed at the local level, such as the Kunstrechnungs Liebende Gesellschaft, founded in Hamburg in 1690, and the Wiskundig Genootschap, founded in Amsterdam in 1778. Since 1860, there had been attempts to establish a mathematical society in France by Michel Chasles and in Germany by Alfred Clebsch. In the latter case, the starting point was the Gesellschaft Deutscher Naturforscher und Ärzte, the Society of German Natural Scientists and Physicians, founded in 1822, which as of 1843 had a special section devoted to astronomy and mathematics.

In 1864, the Moscow Mathematical Society was created. Together with the importance given to personal communication—via the presentation and discussion of mathematical results—was the issuing of a journal devoted to publishing the proceedings of the society and the members' contributions, with the aim of disseminating original research. In the case of the Moscow society, the journal was the *Matematicheskii Sbornik* (first published in 1866). A year later, in 1865, the London Mathematical Society and its official publication, its *Proceedings*, came into existence. This society and its journal served as a model for many others. In particular, it was a model for the Société Mathématique de France and its *Bulletin*, created in 1872; for the Edinburgh Mathematical Society and its journal, created in 1883 and 1884, respectively; for the New York Mathematical Society, created in 1888, which was the forerunner of the American Mathematical Society, created in 1894, and its journal, the *Transactions*, in 1899. Somewhat different was the Circolo Matematico di Palermo, created in 1884 by Giovanni Guccia, a student of Brioschi and Cremona, and its publication, the

Rendiconti, which appeared one year later. One of the differences was the large number of foreign members in the Circolo.

The continued increase in the number of research publications forced mathematicians to engage in another collective enterprise: the creation of review journals, such as the *Jahrbuch über die Fortschritte der Mathematik* (Germany, 1868), the *Repertoire Bibliographique des Sciences Mathématiques* (France, 1885), and the *Revue Semestrielle des Publications Mathématiques* (Amsterdam, 1893). The expansion and specialization of mathematics was the cause of another far-reaching project: the *Encyklopädie der Mathematischen Wissenschaften*, begun in German academies in 1894.

The result of the process that we have summarized above (academic positions, specialized journals, national mathematical societies, review journals) was that, by the end of the nineteenth century, mathematicians were much more professionalized and specialized; mathematical research had become a highly structured activity. Many of the institutions and values of today's mathematical activity were shaped then. This new shape made mathematics much more international.

We have left aside the formation of the German mathematical society. By the last quarter of the century, Germany and the German model had achieved European hegemony in mathematics. We have seen that early attempts to create a mathematical society in Germany failed (the only positive outcome being the founding, in 1868, of the journal *Mathematische Annalen*). The next attempt occurred in the 1880s and succeeded in creating the Deutsche Mathematiker-Vereinigung in 1890. The driving force behind this success was Georg Cantor from Halle (who became the first president of the society). He was supported by Felix Klein from Göttingen. Cantor needed a free professional forum, where he could avoid the hostility from some of his German colleagues. Klein was interested in expanding his standards for the teaching, research, and

Participants in the 1893 International Mathematical Congress held in Chicago. In the center of the front row, Felix Klein. (Courtesy of the American Mathematical Society.)

World's Columbian Exposition in Chicago, 1893.

organization of mathematics. Both Cantor and Klein, each in a different manner, were instrumental in the last steps towards the internationalization of mathematics.

On the occasion of the World's Columbian Exposition in Chicago in 1893, a meeting was arranged by the mathematics faculty of the newly created (1892) University of Chicago. The organizers of the congress were Oskar Bolza, Heinrich Maschke, and Eliakim Hastings Moore, of the University of Chicago, and Henry S. White, of Northwestern University. The congress was held from August 21 to 26. Frank N. Cole reported to the journal *Science* that "the Congress was decidedly cosmopolitan in the authorship of the papers presented." There were 44 papers were presented: 19 came from Germany, 14 from the United States, four from France, three from Italy, two from Austria, one from Switzerland, and one from Russia. Papers were presented by Charles Hermite, David Hilbert, Felix Klein, Hermann Minkowski, Max Noether, Salvatore Pincherle, Alfred Pringsheim, Arthur Schönflies, Edward Study, and others. However, most of the papers were read in the absence of their authors. Attendance was limited: 45 people—three were women. Of the 45, all but four were from the United States. The proceedings of the congress were published as the first volume of the series Papers Published by the American Mathematical Society.

The main figure of the congress was Felix Klein, attending as Imperial Commissioner of the Prussian Ministry of Culture, who brought with him most of the papers by German authors not present at the meeting. Bolza and Maschke were Klein's German students, and Moore had visited Göttingen. Klein gave the opening address, entitled "The Present State of Mathematics," in which he presented an internationalist program:

> A distinction between the present and the earlier period lies evidently in this: that what was formerly begun by a single master-mind, we now must seek to ac-

complish by united efforts and cooperation. A movement in this direction was started in France some time since by the powerful influence of Poincaré. For similar purposes we three years ago founded in Germany a mathematical society, and I greet the young society in New York and its *Bulletin* as being in harmony with our aspirations. But our mathematicians must go further still. They must form international unions, and I trust that this present World's Congress at Chicago will be a step in that direction.

(It is interesting to note in this proclamation the echoes of the renowned internationalist lemma of Karl Marx in the *Communist Manifesto*.)

Papers Published by the American Mathematical Society.—Vol. I.

MATHEMATICAL PAPERS

READ AT THE

INTERNATIONAL MATHEMATICAL CONGRESS

HELD IN CONNECTION WITH THE

WORLD'S COLUMBIAN EXPOSITION CHICAGO 1893

EDITED BY THE
COMMITTEE OF THE CONGRESS

E. HASTINGS MOORE

OSKAR BOLZA   HEINRICH MASCHKE   HENRY S. WHITE

NEW YORK
MACMILLAN AND CO.
FOR THE
AMERICAN MATHEMATICAL SOCIETY
1896

Papers read at the Chicago 1893 International Mathematical Congress. (Courtesy of the American Mathematical Society.)

After the congress, Klein gave a series of lectures at Northwestern University. In the United States, he saw a tremendous opportunity for German mathematics, and he was convinced of the importance of international cooperation in mathematics.

Georg Cantor (1845–1918). First president of the Deutsche Mathematiker-Vereinigung. (From *Gesammelte Abhandlungen*, G. Cantor, Springer 1932.)

After the founding of the Deutsche Mathematiker-Vereinigung, Cantor continued to promote the celebration of an international meeting. Here the reasons were similar to those held before: the need of an international arena where he could defend his mathematical ideas, free from censure and unfair criticism. His was a struggle in the name of scientific freedom. He had many international contacts to whom he addressed his proposal to hold the first international congress of mathematicians. Hermite and Poincaré were among those he contacted. In 1894, Charles Laisant and Émile Lemoine, who had previously exchanged letters on this issue with Cantor, publicly suggested the project in the journal *L'Intermédiaire des Mathématiciens*. From Kazan, in Russia, A. Vassilief proposed a *réunion préparatoire* in Kazan in 1894 and a *congrès constituant* in Belgium or Switzerland in 1897, and also proposed the opening of the first international congress in Paris in 1900. The Deutsche Mathematiker-Vereinigung decided to participate should such a meeting be initiated.

The time was ripe for the first international congress.

PART I
EARLY TIMES

Dear Sir!

As you surely may know the idea of an international congress of mathematicians in later times has been the subject of numerous and vivid deliberations among the savants interested in its occurrence. Looking at the excellent results gained in other fields of science by the international understanding, all those who have considered the issue have unanimously remarked how desirable would be such an international reunion. Once the project had started to take shape, based on numerous conversations and correspondence, and after reiteratively weighting the issue of the site, it was, in general, considered adequate that the first attempt should come from a country especially dedicated to developing international relations because of its situation, customs and traditions. Thus, all eyes turned to Switzerland, and in particular to Zurich.

Zurich's mathematicians had no illusions about the difficulties to be encountered. But they could not decline, in the face of this endeavor and the honorable requests they had received. They decided to adopt all necessary arrangements for the future congress and to contribute as much as they possibly could. Thus, with the concourse of mathematicians from other nations, the organizing committee signing this invitation was constituted, to gather in Zurich in 1897 mathematicians from all countries on Earth.

The Congress, to which by this letter the committee is inviting you to participate, dear Sir, would take place in Zurich on the 9, 10 and 11 of August of 1897, at the Confederal Polytechnical School. The committee will communicate you in due time the detailed program, with the request that you confirm your participation. Allow us, however, to point out at this moment that scientific works and administrative issues will be organized, naturally, around those having general interest and importance.

The importance of the scientific congresses depend on the care for the personal relations. The local committee has the task of devoting attention to this feature of the congress, and of taking into account the shaping of a modest festive program.

Hopefully all expectation associated to this first mathematical meeting will be fulfilled! Hopefully the numerous participation will foster scientific and personal relations among colleagues in the interest of the common work and the progress of mathematical sciences!

THIS ELEGANT LETTER was the first invitation letter for an International Congress of Mathematicians. It marks the starting point of the long history of the 25 congresses that have taken place through the last 110 years. The letter reflects all the hopes that were associated with the idea of reuniting mathematicians from countries all around the world. It expresses the expectations for future success.

In this part, we will see the first period of the history of the ICM, that of the "Early Times." It is comprised of the first five congresses, which were held in

- Zurich, August 9–11, 1897;
- Paris, August 6–12, 1900;
- Heidelberg, August 8–13, 1904;
- Rome, April 6–11, 1908;
- Cambridge, August 22–28, 1912.

As we will see, these congresses followed each other at an ever quickening pace. The number of mathematicians attending, the number of countries represented, the number of lectures delivered increased. The scope, aims, and goals of the congresses expanded. Most of

Internationaler Mathematiker-Kongress in Zürich 1897.

Zürich, Februar 1897.

An die Mitglieder des internationalen Komitees.

Hochgeehrte Herren!

Das Lokalkomitee ist mit der Ausarbeitung eines Programmes für die Verhandlungen des Kongresses beschäftigt. Für Montag, den 9., und Mittwoch, den 11. August, sind Gesamtsitzungen vorgesehen, in welchen Fragen von allgemeinerem Interesse behandelt werden sollen. Vorträge speziellerer Natur würden, je nach Bedürfnis in Sektionen, am Dienstag gehalten werden.

Neben rein wissenschaftlichen Fragen soll der Kongress seine Aufmerksamkeit auch Angelegenheiten mehr geschäftlicher Natur zuwenden. Wir rechnen hierzu Fragen der Bibliographie, der Lexikographie, der Terminologie, die Inangriffnahme gemeinsamer wissenschaftlicher Unternehmungen (historische Arbeiten, zusammenfassende Referate, Herausgabe von Werken, Veranstaltung von Ausstellungen) u. dergl. Auch Besprechungen, die sich mit den Beziehungen der Mathematik zu anderen Wissensgebieten, zur Technik, zum öffentlichen Leben etc. beschäftigen, könnten in den Bereich der Verhandlungen aufgenommen werden.

Wir ersuchen Sie nun, uns Ihre Ansichten über die Organisation und die Traktanden des Kongresses mitteilen zu wollen und uns womöglich aus der Reihe jener geschäftlichen Fragen bestimmte Themata zu bezeichnen, über welche Sie entweder selbst zu referieren wünschen oder für welche dann andere Referenten zu bestellen wären.

Empfangen Sie zum voraus für Ihre Bemühungen unsern verbindlichsten Dank.

Mit ausgezeichneter Hochachtung

Das Lokalkomitee.

The invitation letter for the first ICM, Zurich 1897. (Courtesy of the Bildarchiv der ETH-Bibliothek Zürich.)

all, the general excitement and commitment to the idea of an international congress of mathematicians grew tremendously. As we will see, in this first period, all the expectations were fulfilled.

ZURICH 1897

WE HAVE JUST SEEN the first invitation letter issued for an international congress. It was signed by an international group of mathematicians. Although the majority of the members were Swiss, nine different countries were represented: the Austro-Hungarian Empire, France, Great Britain, Germany, Italy, Russia, Sweden, Switzerland, and the United States of America.

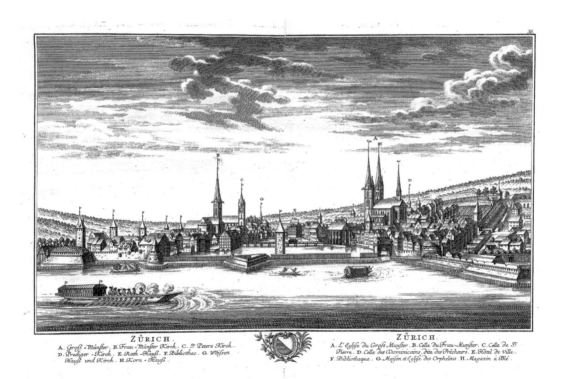

ZÜRICH.
A. Groß-Münster. B. Frau-Münster Kirch. C. St Peters Kirch. D. Prediger-Kirch. E. Rath-Hauß. F. Bibliothec. G. Wäisen Hauß und Kirch. H. Korn-Hauß.

ZÜRICH.
A. L'Eglise du Groß Munster. B. Celle du Frau-Munster. C. Celle de St Pierre. D. Celle des Dominicains, dite des Prêcheurs. E. Hôtel de Ville. F. Bibliotheque. G. Maison et Eglise des Orphelins. H. Magasin à Blé.

Zurich in the eighteenth century. (Courtesy of the Universität Bern, "Sammlung Ryhiner.")

Mathematicians signing the invitation letter for the first ICM in Zurich in 1897. (From the proceedings of the 1897 ICM, Teubner 1898.)

The letter was sent from Zurich in January 1897. Perhaps because of the newness of the societies, or for some other more intricate reason, it was decided not to send invitations through the mathematical societies but directly to individual mathematicians. The addressees were 2000 mathematicians and physicists, who received either the German or the French version of the letter. (There were slight differences between the two versions; for example, in the French one, "intimate meetings" was added to the social program.) It is also interesting to see how the letter was delivered. It was not sent from Zurich but distributed by mathematicians representing the organizing committee in the cities of Woolwich (Great Britain), Palermo (Italy), West Nyack (U.S.A.), Göttingen (Germany), Paris (France), Gent (Belgium), St. Petersburg (Russia), Vi-

The first reception of the 1897 congress took place at the Tonhalle, Zurich's concert hall. (Courtesy of Historic Print & Map Company, http://www.ushistoricalarchive.com/.)

enna (Austria), Stockholm (Sweden), Groningen (Holland), Athens (Greece), and Port (Portugal). Many of the mathematical journals also helped to send the invitation.

This was the first success of the congress, a careful and thorough involvement of a large number of mathematicians from different countries.

After a lengthy preparation, the first congress finally began. On the evening of Sunday, August 8, the participants were officially welcomed at the Tonhalle, Zurich's concert hall. There, after some "friendly chatting," Adolf Hurwitz, head of the reception committee, addressed the group with these words:

May the inspiring force of personal communication rise during these days, providing plenty of occasions for scientific discussions. May we together enjoy the relaxed and cheerful comradeship, enhanced by the feeling that here representatives of many different countries feel united by the most ideal interests in peace and friendship.

This short paragraph has become a symbol of future international congresses. It expresses very clearly both the scientific and the human intention of the congresses.

The next day, the congress was officially inaugurated at the Aula of the Eidgenössische Polytechnikum (the Confederal Polytechnical School) (see page 155). Karl F. Geiser, president of the organizing committee, in a long and florid address praised the Swiss mathematical glories: Jacob, Johann, and Daniel Bernoulli, Leonhard Euler, and Jakob Steiner.

This was the first meeting where mathematicians from different countries met together, except for the Chicago 1893 meeting, to which we referred before. Here, there were no precedents to follow. To overcome this difficulty, the organizing committee established some "Regulations for the Congress" (we will return later to these regulations). The first task was to approve these regulations. Once this was accomplished,

they were immediately applied, and Geiser was elected president of the congress "by acclamation."

A similar situation occurred with the scientific program of the congress. There were no precedents to follow. Thus, it was proposed to follow the model of the congresses "with itinerating venue" of the Schweizerischen Naturforschenden Gesellschaft (the Swiss Society of Nature Scientists), where there were plenary sessions with lectures of general interest given by lecturers chosen by invitation, and there were also specialized sections. In these sections, lecturers were kindly requested to speak for no more than thirty minutes. This general structure has been retained to the present. Present day difficulties are still the same: it was reported that one of the lectures in the section on Mechanics and Mathematical Physics was not delivered due to the confusion of repeated changes of lecture room.

At the first plenary session, there were two plenary lectures:

"Sur les rapports de l'analyse pure et de la physique mathématique," by Henri Poincaré from Paris;

"Über dieEntwickelung der allgemeinen Theorie der analytischen Funktionen in neuerer Zeit," by Adolf Hurwitz from Zurich.

Adolf Hurwitz (1849–1919), plenary lecturer in 1897. (Courtesy of G. L. Alexanderson.)

Henri Poincaré (1854–1912), plenary lecturer in 1897. (Courtesy of the Archives Henri Poincaré.)

Sur la théorie des nombres premiers.

Par

Ch. de la Vallée Poussin à Louvain.

M. de la Vallée Poussin s'est occupé de la fréquence des nombres premiers de différentes formes dans un Mémoire étendu, publié dans les Annales de la Société scientifique de Bruxelles (1896) sous le titre: Recherches analytiques sur la théorie des nombres premiers.
Voici quelques conclusions de ce travail, concernant les nombres premiers d'une forme linéaire primitive $Mx + N$:
1° L'expression

$$\frac{\varphi(M)}{y} \sum_{N < y} lq_x,$$

dans laquelle la somme est étendue aux nombres premiers $< y$ et de la forme $Mx + N$, a pour limite l'unité quand y tend vers l'infini.
2° La différence

$$\varphi(M) \sum_{N < y} \frac{lq_N}{q_N} - ly$$

tend vers une limite finie et déterminée quand y tend vers l'infini.
3° Le nombre des nombres premiers de la forme $Mx + N$ et $< y$ peut se représenter par l'expression

$$\frac{1 + \varepsilon}{\varphi(M)} \frac{y}{ly}$$

où ε tend vers zéro quand y tend vers l'infini.

Lecture by Charles de la Vallée Poussin in 1897 in Zurich. (From the proceedings of the 1897 ICM, Teubner 1898.)

Since Poincaré was not able to attend the congress, "due to a death" as cryptically explained in the proceedings, his manuscript was read to the congress by Jerôme Franel. The afternoon of this first congress day was devoted to leisure.

The next day, Tuesday, August 10, was devoted to the lectures organized in the five scientific sections of the congress:

* Section I: Arithmetic and Algebra,
* Section II: Analysis and Function Theory,
* Section III: Geometry,
* Section IV: Mechanics and Mathematical Physics,
* Section V: History and Bibliography.

Among the lectures scheduled in the sections, we find *"Sur la théorie des nombres premiers"* by the young Belgian mathematician Charles de la Vallée Poussin, who had just a year before proven the Prime Number Theorem (independently and at the same time as Jacques Hadamard). Unfortunately for the audience,

the of time prevented de la Vallée Poussin from delivering his lecture.

Felix Klein (1849–1925), plenary lecturer in 1897. (Courtesy of G. L. Alexanderson.)

Giuseppe Peano (1858–1932), plenary lecturer in 1897. (Courtesy of the Dipartimento di Matematica dell' Università degli Studi di Torino: *Fonti iconografiche* della Biblioteca Matematica "Giuseppe Peano.")

In the session of Wednesday, August 11, the other two plenary lectures were delivered:

* *"Logica matematica,"* by Giuseppe Peano from Turin;
* *"Zur Frage des höheren mathematischen Unterrichtes,"* by Felix Klein from Göttingen.

In this last session, it was discussed and decided that the successive congresses should follow at intervals of three to five years and that the venue of the following congress should be decided at the end of each meeting. The next congress was to be in 1900 in Paris, organized by the Société Mathématique de France.

The congress was concluded by its president, Geiser, with a cheerful "Auf Wiedersehn in Paris!"

As reported by George Bruce Halsted from Austin, Texas, to the journals *Science* and *The American Mathematical Monthly*, the congress "was in every way a success." The general feeling of the participants was expressed by Émile Picard, president of the Société Mathématique de France, at the closing banquet: "The

success of our fist meeting is a warrant for the future of the institutions just founded."

This success was twofold. On one hand, it was a success in attendance and in the scientific level. Four plenary lectures and 30 section lectures were presented. Many of the most eminent mathematicians of the time attended, including Bendixson, Borel, Brunn, Cantor, Fredholm, Hausdorff, Hobson, Levi-Civita, Lindelöf, Mellin, Mertens, Mittag-Leffler, Minkowski, M. Noether, Padé, Pincherle, Pringsheim, Schönflies, Segre, Tauber, and Volterra. There was only one boycott to the congress: no member of the University of Berlin attended. The role of strong personalities in the German academic world had always been very important, and this congress was, in a sense, a congress of the Göttingen group, with Felix Klein as the leading figure. The total attendance rose to 208 mathematicians, from 16 countries. It is noteworthy

Nach Ländern geordnet ergiebt diese Teilnehmerliste die folgende Gruppierung:

Land.	Herren.	Damen.
Schweiz	60	8
Deutschland	41	12
Frankreich	23	6
Italien	20	5
Oesterreich-Ungarn	17	3
Rufsland	12	1
Nordamerika	6	1
Schweden	6	—
Finland	4	1
Belgien	3	—
Dänemark	3	—
Grofsbritannien	3	—
Holland	3	—
Spanien	1	1
Griechenland	1	—
Portugal	1	—
	204	38

Im ganzen waren also 16 Länder durch 242 Teilnehmer vertreten, worunter 38 Damen.

Distribution of members of the 1897 congress according to their nationality. (From the proceedings of the 1897 ICM, Teubner 1898.)

that four participants were women: Dr. Fräulein Iginia Massarini from Rome; Professor Frau Vera von Schiff from St. Petersburg; Professor Fräulein Charlotte Angas Scott (who was the first British woman to receive a doctorate in mathematics) from Bryn Mawr, Pennsylvania; and Dr. Frau Charlotte Wedell from Göttingen. Hurwitz did not overlook this presence when he started his address with *"Kolleginen und Kollegen!"*

The other success of the congress was the atmosphere. This is very well illustrated by the Regulations that were approved. The first article established the objectives of the congress; the first two objectives were

a) to foster personal relations between mathematicians of different countries;

b) to present in the lectures of the plenary sessions and the different sections an overview of the current state of the different areas of mathematical sciences and their applications, and to discuss specific problems of particular importance.

The order in which these two aims were presented is especially noteworthy. It reveals the importance for the development of mathematics assigned to the personal relations between mathematicians. The joy that marked all the social activity of the congress reflects this desire. From the very beginning, the congress was able to foster a very pleasant atmosphere. For example, no requirements were placed on dress for scientific sessions or banquets (remember that the congress took place in the nineteenth century). We will look at this aspect of the congress in detail later in "Social Life at the ICM."

Part of the pleasant atmosphere was due to the delicate dealing with all issues related to the balance of nationalities. It was not by chance that the decision was made to hold the congress in Zurich, "at the crossroad of the large railways from Paris to Vienna and from Berlin to Rome," as Geiser explained at the opening ceremony. This national equilibrium was also seen in the language issue. Of the seven articles of the Regulations of the Congress, two were devoted to language, stating that the official publications of the congress would be in French and German, and the proceedings would have one edition in each language. In public speeches and voting, English and Italian were also allowed (in any case, the number of English-speaking participants was very small: the lecture by E. Schröder from Karlsruhe initially to be delivered in English was finally delivered in German due to the small number of English-speaking participants attending the lecture). The choice of the plenary lecturers also reveals a careful balance between nationalities: one Frenchman, one German, one Italian, and Adolf Hurwitz, who, although German, was from the host institution in Zurich. The only shadow of the German–French rivalry appeared after the approval of Paris as the venue of the 1900 congress, when Felix Klein, at that time president of the Deutsche Mathematiker-Verenigung, immediately declared the great interest of his society in organizing the 1904 congress.

The feeling that mathematics was entering a new era based on international cooperation was very strong. Ferdinand Rudio, one of the Swiss organizers of the congress, listed the new directions opened by the congress, which needed international agreements. They included the issue of unifying mathematical terminology and units, as had been done recently by the physicists with the volt, the ampere, and the ohm. There was also the need for an international literary journal for mathematics; the *Jahrbuch über die Fortschritte der Mathematik* had been in publication since 1868, and the *Repertoire Bibliographique des Sciences Mathématiques* since 1885, but they were slow in delivering the information on the developments of a science producing many more results at a much faster speed. He also mentioned the need for a general classification of mathematics that would help the bibliographic effort. He pointed out that the list of

Ferdinand Rudio proposed at the Zurich 1897 congress to publish Euler's collected works. (Courtesy of the Bildarchiv der ETH-Bibliothek Zürich.)

participants of the congress should be seen as the starting point of an international directory of mathematicians, where one could find addresses and field of speciality of all mathematicians in the world (this idea had to wait until 1958 when the first *World Directory of Mathematicians* appeared). Rudio complemented this idea with the proposal of creating a biographical dictionary of current mathematicians, which would include portraits of the most important. Rudio summarized all projects in a lemma: *"Viribus unitis! sei unsere Losung"* ("United our forces! This is our watchword").

Let us end this account of the first International Congress by focusing on three features that will repeatedly recur when viewing congresses to come. One is the particular attention paid to the past of mathematics (beyond the always present section on the History of Mathematics). This attention has appeared in the interest of rendering tribute to glorious mathematicians, present and past; in the publication of their works; and in the honoring of their memory. In this first

"United our forces! This is our watchword." (From the *Encyklopädie der Mathematischen Wissenschaften.*)

Zurich congress, Geiser in his opening speech praised five eminent Swiss mathematicians. At a proposal of Francesco Brioschi from Milan and Gösta Mittag-Leffler from Stockholm, there was a toast in honor of Charles Hermite, and a telegram was sent to him.

Another recurring feature of the ICM is the presentation of books. Rudio proposed a motion requesting the publication of the complete works of Leonhard Euler. Andrey Markov from St. Petersburg had announced that he would give a lecture entitled "On the Edition of the Works of Tschebyschef," but in the end he did not attend the congress. It is also worth noting the presence, as members of the congress, of three publishers: Alfred Ackermann-Teubner from Leipzig; Albert Gauthier-Villars from Paris; and Ulrico Hoepli from Milan.

The third aspect of the ICM that we will be considering is the financing of the congresses. While not a great deal of information is available, we do know that the congress expenses were paid from subsidies from confederal, cantonal, and local governments. There were also donations from many local businessmen, and there was a registration fee per participant of 25 Swiss francs.

PARIS 1900

Who of us would not be glad to lift the veil behind which the future lies hidden; to cast a glance at the next advances of our science and at the secrets of its development during future centuries? What particular goals will there be towards which the leading mathematical spirits of coming generations will strive? What new methods and new facts in the wide and rich field of mathematical thought will the new centuries disclose?

History teaches the continuity of the development of science. We know that every age has its own problems, which the following age either solves or casts aside as profitless and replaces by new ones. If we would obtain an idea of the probable development of mathematical knowledge in the immediate future, we must let the unsettled questions pass before our minds and look over the problems which the science of today sets and whose solution we expect from the future. To such review of problems the present day, lying at the meeting of the centuries, seems to me well adapted. For the close of a great epoch not only invites us to look back into the past but also directs our thoughts to the unknown future.

W ITH THESE WORDS David Hilbert from Göttingen started his lecture on Wednesday, August 8, 1900, at nine in the morning in the Chasles Amphitheater of the Faculty of Science of the Sorbonne in Paris (named after the geometer Michel Chasles). This lecture has become the most recognizable icon of the ICM. In it, Hilbert presented a list of 23 problems, with the intention of illuminating (and perhaps also de-

termining) the future of mathematics in the twentieth century.

SUR LES

PROBLÈMES FUTURS DES MATHÉMATIQUES,

Par M. David HILBERT (Göttingen),

TRADUITE PAR M. L. LAUGEL. (¹)

Qui ne soulèverait volontiers le voile qui nous cache l'avenir afin de jeter un coup d'œil sur les progrès de notre Science et les secrets de son développement ultérieur durant les siècles futurs? Dans ce champ si fécond et si vaste de la Science mathématique, quels seront les buts particuliers que tenteront d'atteindre les guides de la pensée mathématique des générations futures? Quelles seront, dans ce champ, les nouvelles vérités et les nouvelles méthodes découvertes par le siècle qui commence?

L'histoire enseigne la continuité du développement de la Science. Nous savons que chaque époque a ses problèmes que l'époque suivante résout, ou laisse de côté comme stériles, en les remplaçant par d'autres. Si nous désirons nous figurer le développement présumable de la Science mathématique dans un avenir prochain, nous devons repasser dans notre esprit les questions pendantes et porter notre attention sur les problèmes posés actuellement et dont nous attendons de l'avenir la résolution. Le moment présent, au seuil du vingtième siècle, me semble bien choisi pour passer en revue ces problèmes; en effet, les grandes divisions du

(¹) L'original de la traduction a paru en allemand dans les *Göttinger Nachrichten*, 1900. M. Hilbert a fait ici quelques modifications à l'original au § 13 et quelques additions au § 14 et au § 23. (L. L.)

Hilbert's lecture as it appeared in the proceedings of the Paris 1900 congress. (From the proceedings of the 1900 ICM, Gauthier-Villars 1902.)

But Hilbert's lecture was not a plenary one. How could this be, since at the time Hilbert was regarded as a leading mathematician of his generation? Hilbert was, indeed, invited to give a plenary address to the congress. However, he delayed choosing his topic, consulting with his friends Hermann Minkowski and Adolf Hurwitz. Finally he decided to discuss a list of problems. The writing of the paper was delayed so long that the title of his lecture was left out of the program of the congress. Once in Paris, Hilbert had to deliver the lecture in the joint session of the sections on Bibliography and History, and Teaching and Methods, presided over by the German historian of mathematics Moritz Cantor.

In the long preamble of the lecture, Hilbert discussed the nature of problems in mathematics and their role in the advance of the science. Before starting to discuss the list of problems, Hilbert expressed his faith that "In mathematics there is no *ignorabimus*." This statement was in opposition to the famous aphorism of the German physiologist Emil du Bois-Reymond (whose brother Paul was a well-known mathematician): *ignoramus et ignorabimus* (we do not know and will not know).

Due to limited time, Hilbert presented only ten of the 23 problems:

1. Cantor's Problem of the Cardinal Number of the Continuum,

2. The Compatibility of the Arithmetical Axioms,

6. Mathematical Treatment of the Axioms of Physics,

8. Problems of Prime Numbers,

12. Extension of Kronecker's Theorem on Abelian Fields to any Algebraic Realm of Rationality,

13. Impossibility of the Solution of the General Equation of the 7th Degree by means of Functions of only Two Arguments,

SUR LES

PROBLÈMES FUTURS DES MATHÉMATIQUES,

PAR M. DAVID HILBERT (Göttingen),

I. — Problème de M. Cantor relatif à la puissance du continu.

II. — De la non-contradiction des axiomes de l'Arithmétique.

III. — De l'égalité en volume de deux tétraèdres de bases et de hauteurs égales.

IV. — Problème de la ligne droite. plus court chemin d'un point à un autre.

V. — De la notion des groupes continus de transformations de Lie, en faisant abstraction de l'hypothèse que les fonctions définissant les groupes sont susceptibles de différentiation.

VI. — Le traitement mathématique des axiomes de la Physique.

VII. — Irrationalité et transcendance de certains nombres.

VIII. — Problèmes sur les nombres premiers.

IX. — Démonstration de la loi de réciprocité la plus générale dans un corps de nombres quelconque.

X. — De la possibilité de résoudre une équation de Diophante.

XI. — Des formes quadratiques à coefficients algébriques quelconques.

XII. — Extension du théorème de Kronecker sur les corps abéliens à un domaine de rationalité algébrique quelconque.

XIII. — Impossibilité de la résolution de l'équation générale du septième degré au moyen de fonctions de deux arguments seulement.

XIV. — Démontrer que certains systèmes de fonctions sont finis.

XV. — Établissement rigoureux de la Géométrie énumérative de Schubert.

XVI. — Problèmes de topologie des courbes et des surfaces algébriques.

XVII. — Représentation des formes définies par des sommes de carrés.

XVIII. — Partition de l'espace en polyèdres congruents.

XIX. — Les solutions des problèmes réguliers du calcul des variations sont-elles nécessairement analytiques?

XX. — Problème de Dirichlet dans le cas général.

XXI. — Démonstration de l'existence d'équations différentielles linéaires ayant un groupe de monodromie assigné.

XXII. — Relations analytiques exprimées d'une manière uniforme au moyen de fonctions automorphes.

XXIII. — Extension des méthodes du Calcul des variations.

The 23 problems of Hilbert. (From the proceedings of the 1900 ICM, Gauthier-Villars 1902.)

16. Problems of the Topology of Algebraic Curves and Surfaces,

19. Are the Solutions of Regular Problems in the Calculus of Variations Always Necessarily Analytic?

21. Proof of the Existence of Linear Differential Equations Having a Prescribed Monodromic Group,

22. Uniformization of Analytic Relations by Means of Automorphic Functions.

David Hilbert (1862–1943). (Courtesy of G. L. Alexanderson.)

The numbers correspond to those of the paper published later. The lecture was delivered in German (in the clear Prussian German spoken by Hilbert), but to help the audience understand, an abridgment in French was distributed before the talk. It was entitled *Sur les problèmes futurs des Mathématiques*. The printed version of the whole paper was entitled *Mathematische Probleme* (Mathematical Problems).

The fame of these problems has increased with time, giving prestige to any contribution related to their solution. But how was the lecture received then? The answer is unclear from the reports of the time. The chronicle by George Bruce Halsted for *The American Mathematical Monthly* referred to "Hilbert's beautiful paper," while Charlotte Angas Scott reporting for the *Bulletin of the American Mathematical Society* commented on the "rather desultory discussion that

followed." When reporting to his friends about the congress, Hilbert did not seem particularly satisfied.

Hilbert's lecture has overshadowed the rest of the Paris congress. The beginning of the new century was celebrated in Paris with an *Exposition Universelle*, a World's Fair, in the style of large international exhibitions that had been taking place since the mid-nineteenth century (see page 22). The event lasted for more than six months and convened many other scientific meetings in the city; more than 200 congresses were held in Paris that year in connection with the *Exposition Universelle*.

PREMIÈRE PARTIE. — DOCUMENTS ET PROCÈS-VERBAUX.   21

SECTIONS V ET VI. — BIBLIOGRAPHIE ET HISTOIRE.
ENSEIGNEMENT ET METHODES.

Mercredi 8 août.
Présidence de M. M. CANTOR.

Première séance.

En raison de l'absence du Président de la cinquième Section, M. le prince Roland Bonaparte, les Sections V et VI se réunissent sous la présidence de M. Cantor, Président de la sixième Section. MM. d'Ocagne et Laisant remplissent les fonctions de Secrétaires.

La séance est ouverte à 9ʰ.
Communications :

1. D. HILBERT, *Sur les problèmes futurs des Mathématiques.*
M. Peano déclare que la Communication ultérieure de M. Padoa répondra au problème n° 2 de M. Hilbert. M. Mehmke rappelle qu'il a proposé certaines représentations monographiques dans l'espace d'où pourrait résulter une solution de l'équation générale du septième degré.

2. R. FUJISAWA, *Note on the Mathematics of the old Japanese school.*
M. A. Vassilief demande si l'on ne peut pas trouver les traces de l'influence grecque, par l'intermédiaire du royaume gréco-bactrien, sur les premiers géomètres japonais.

3. LEAU, *Proposition d'un vœu pour l'adoption d'une langue scientifique universelle.*
Le vœu que M. Leau propose au Congrès d'émettre est le suivant :
1° *Il y a lieu d'adopter une langue scientifique et commerciale universelle.*
2° *Les Académies officielles sont respectueusement invitées à s'entendre pour la réalisation de ce projet.*
M. Leau propose, en outre, la résolution suivante :
Le Congrès décide de nommer cinq Membres à la Délégation qui se forme :
1° *Pour demander aux Académies officielles de vouloir bien adopter une langue auxiliaire universelle;*

Hilbert's lecture was presented in the Section on Bibliography and History of the Paris 1900 congress. (From the proceedings of the 1900 ICM, Gauthier-Villars 1902.)

Paris during the Exposition Universelle in 1900. (From *Exposition Universelle de 1900*, Resengoti 1900.)

ticipants, "with the attendance of numerous dames and ladies splendidly dressed," were very happy.

First, the formal part of the congress began. At the proposal of the French philosopher of mathematics Jules Tannery, representing the French Ministry of Instruction, Henri Poincaré was named (by acclamation) president of the congress, and, at Poincaré's proposal, Charles Hermite was named honorary president. Poincaré explained that Hermite, due to his advanced age, could not attend the congress but he "is in heart with us."

Charles Hermite (1822–1901), honorary president of the Paris 1900 congress. (From *The Mathematician Sophus Lie* by Arild Stubhaug, Springer 2002.)

It was expected that the exhibition would attract people to the congress; more than 1000 mathematicians and almost 700 family members declared their intention of attending before December 1899. Unfortunately, the opposite occurred. The tremendous heat of July, newspapers reporting on "exhibition crowds and exhibition extortions" (*Nature*, 1900), and rumors about the difficulties in obtaining accommodation caused attendance to be just a quarter higher than that of Zurich in 1897.

The opening session of the 1900 International Congress took place in the Palais des Congrès on the grounds of the *Exposition Universelle* on Monday, August 6, at nine in the morning (see page 157). The site was magnificent; the weather was perfect and par-

Next the naming of the congress officials took place: ten vice-presidents; one secretary general; five secretaries; six section presidents; and another six section secretaries. Finally, the congress could commence, and it did with two plenary lectures:

* *"L'historiographie des mathématiques,"* by Moritz Cantor from Heidelberg;

"Betti, Brioschi, Casorati—Trois analystes italiens et trois manières d'envisager les questions d'analyse," by Vito Volterra from Rome.

Both lectures were delivered in French. Language turned out to be an issue at this congress, but in this case because of the absolute preponderance of French. At eleven-thirty that morning, the session was adjourned. The afternoon was free, and many congress attendants devoted time to discovering the attractions offered by the *Exposition*.

The next day, the activity of the sections commenced at the Sorbonne, where the Rector of the Académie de Paris had graciously offered the congress the use of three amphitheaters of the Faculty of Science named after French mathematicians or astronomers: the Cauchy Amphitheater for Sections I (Arithmetic and Algebra) and III (Geometry); the Le Verrier Amphitheater for Sections II (Analysis) and IV (Mechanics); and the Chasles Amphitheater for Sections V (Bibliography and History) and VI (Teaching and Methods). (See page 156.) This last section was the only new addition to the Zurich scheme of sections for the international congresses. After the session, there was a lunch at the École Normale Supérieure where "a pleasant opportunity for social intercourse was enjoyed."

Charlotte Angas Scott, in her 23-page report of the congress, spoke of 200 people attending the plenary lectures and 90 the sectional meetings, organized in two parallel sessions. However, the schedule of the sections seems to have been relaxed. Indeed, Section I, for example, met on Tuesday, August 7, from nine to eleven, where four communications were presented; on Thursday it met again from nine to ten thirty and only two of the five scheduled communications were presented (this was due to the absence of three of the lecturers); and the meeting for Friday was canceled. The attractions of the *Exposition* were too appealing!

Among odd events, let us recall the discussion that took place in Sections V and VI, combined because of the absence of the president of Section V, Prince Roland Bonaparte. It was proposed at the first meeting, "the adoption of a universal scientific and commercial language," like Esperanto (which was a fairly recent creation). There was little sympathy for this proposal, and the discussion extended to the next meeting of the sections. In any case, the problem was seen as a challenge. A. Vassilief explained that, in the beginning of the nineteenth century, it was sufficient for a scholar to know three languages—Latin, English, and French—and he warned that 20 or 30 scientific languages would be a great danger for science. A final recommendation was adopted with the approval of the majority of the congress: "that the Academies and learned Societies from all countries study the proper means to remedy the harms coming from the increasing diversity of languages employed in the scientific literature."

Prince Roland Bonaparte was a peculiar character. A grandnephew of Napoleon Bonaparte, he was well known at the time for his generous commitment to all types of scientific enterprises, mostly those related to natural history. The report for the scientific journal *Nature* explains that "[a] fête had been organized by President [of the Republic] Loubet for Thursday evening, but could not be held on account of the funeral of the King of Italy; the invitations were transferred to the fête in honor of the Shah," which was a "scientific soirée" organized at his estates by Prince Roland Bonaparte, where the members of the ICM met with their colleagues of the Physics Congress. Halsted reported that the evening was a "delightful entertainment."

The closing ceremony took place presided over by Poincaré in the magnificent Richelieu Amphitheater of the Sorbonne (see page 24). The decision was taken to entrust the Deutsche Mathematiker-Vereinigung with the organization and the choice of venue of the next

The Richelieu Amphitheater, where the closing session of the Paris 1900 congress took place. (© Olivier Jacquet—Université Paris-Sorbonne.)

congress in 1904. The site, at that moment, seemed to be the thermal city of Baden-Baden.

The session ended with the plenary lectures (both in French):

* *"Du rôle de l'intuition et de la logique en mathématiques,"* by Henri Poincaré;
* *"Une page de la vie de Weierstrass,"* by Gösta Mittag-Leffler from Stockholm.

Charlotte Angas Scott reported bluntly on the presentation of papers in the congress, which in her opinion were "usually shockingly bad" since "instead of speaking to the audience, [the lecturer] reads his paper to himself in a monotone that is sometimes hurried, sometimes hesitating, and frequently bored ... so that he is often tedious and incomprehensible."

Moritz Cantor (1829–1920), plenary lecturer in 1900. (Courtesy of the Universitätsarchiv Heidelberg.)

Gösta Mittag-Leffler (1846–1927), plenary lecturer in 1900. (Courtesy of G. L. Alexanderson.)

The only one spared this devastating criticism was Mittag-Leffler, whose presentation she described as "admirable and engaging a style . . . It is not given to everyone to do it with this charm."

Henri Poincaré (1854–1912), plenary lecturer in 1900. (Courtesy of the Archives Henri Poincaré.)

Vito Volterra (1860–1940), plenary lecturer in 1900. (Courtesy of the Archivio Storico dell' Accademia Nazionale delle Scienze detta dei XL.)

The day after the closing of the congress, a banquet was held at noon in the Salle de l'Athénée-Saint-Germain. About 160 members attended. During the toasts, Gaston Darboux apologized for Poincaré, who was "too tired to be able to participate."

Let us give some figures to compare this congress with that of Zurich. As to attendance, this congress represented an improvement: from 208 participants in Zurich the number rose to 250, and the number of countries represented from 16 to 26. The largest national group was, naturally, the French with 95 participants, followed by the Germans with 26, Italians 23, North Americans 19, Russians 14 (which included the two Lindelöf brothers from the Duchy of Finland), British 12, Belgians 12, and other countries with smaller figures. The number of women mathematicians attending also increased from four in Zurich to six in Paris. They were, as reported in the proceedings, Madame Ely Achsale from Poughkeepsie, New York; Madame A. Jolles from Berlin; Madame Olga Sabine from Moscow; Mademoiselle Elna Sarauw from Copenhagen; Madame Vera Schiff from St. Petersburg; and Miss Charlotte Angas Scott from Bryn Mawr, Pennsylvania. Two of these women, Schiff and Scott, had already attended the Zurich congress. (In 1905 Scott became vice president of the American Mathematical Society.) There was even an increase in the number of publishers attending the congress, from three to four: Ackermann-Teubner from Leipzig; Gauthier-Villars from Paris; Hermann from Paris; and C. Naud from Paris. Without precise information of exchange rates and comparative inflation, the registration fee for the Paris congress seems not too dissimilar to that of Zurich: 30 francs for the participant and five for every family member.

Other than figures, a comparison with the Zurich congress is not satisfactory. There was a great deal of criticism of the arrangements of the congress: after the arrival in Paris, obtaining information was very difficult for participants (the opening was scheduled for two-thirty in the afternoon and then changed to nine-thirty in the morning the same day, causing many congress members to be late); the lack of a common assembly room for the congress was also a serious defect. Clearly the coincidence with the *Exposition Universelle* caused a great many of these difficulties, as did, in the opinion of C. A. Scott, entrusting part of the organization of the congress to a private firm. But more importantly, there was dissatisfaction with the lack of action to complete the tasks that the Zurich congress had set regarding terminology, bibliography, classification of mathematics, an international directory, and other concerns.

HEIDELBERG 1904

His Royal Highness the Great Archduke heir Friedrich von Baden has gracefully accepted the Honorary Presidency of the III International Congress of Mathematicians and has commissioned me to be in charge of the opening. Thereby, I warmly welcome all who have come attending to our invitation.

WITH THESE POMPOUS WORDS Heinrich Weber from Strasbourg (a German city at that time) opened the Heidelberg congress on August 9, 1904. This time the congress had a special character, as the Deutsche Mathematiker-Verenigung had decided to link the congress with the official commemoration of the centenary of the birth of the great German mathematician Carl Gustav Jacob Jacobi (1804–1851). Hans Amandus Schwarz from Berlin closed the opening session saying:

> It has corresponded to me the high honor of expressing to the German Mathematical Society our gratitude for

Lithography included in the proceedings of the 1904 congress showing Heidelberg, the Castle, and the Neckar River. (From the proceedings of the 1904 ICM, Teubner 1905.)

The opening ceremony of the 1904 congress took place in the Heidelberg Museum, then part of the university. (Courtesy of the Universitätsarchiv Heidelberg.)

the invitation to participate in the commemoration for Jacobi envisioned at this congress. In the name of the Royal Prussian Academy of Sciences, and the Rector and Senate of the Royal University Friedrich Wilhelm of Berlin, and the representatives here present from the Royal University Albertus of Königsberg in Prussia we thank you.

It is evident as we go into detail that this congress showed the splendor of the German Empire.

The organizing committee set up for this congress was fairly large (23 members, from 19 different institutions, including the publisher A. Ackermann-Teubner), but it was formed exclusively of German members. A great effort to promote attendance had been made by sending the invitation letter to 2000 mathematicians from all over the world. All members of the main mathematical societies (the Deutsche Mathematiker-Vereinigung, the Société Mathématique de France,

the London Mathematical Society, the Wiskundig Genootschap te Amsterdam, the Circolo Matematico di Palermo, the mathematical societies of Moscow and Kazan, and the American Mathematical Society) were invited. Mathematicians from Hungary, Sweden, Norway, Spain, Portugal, and other countries supplied lists of addresses. The invitation was also included in more than 25,000 issues of the main mathematical journals. All mathematical publications by the company B. G. Teubner included (at no cost) a short note of invitation. This effort paid off: participation went to 336, 128 more participants than in 1897 in Zurich, and 86 more than in 1900 in Paris.

The opening ceremony took place in the meeting room of Heidelberg's Museum, part of the University of Heidelberg. Once elected president of the congress, Weber spoke an elegant tribute to the memory of past mathematicians. He recalled a long list of mathematicians who had recently died: Karl Weierstrass, in 1897; Charles Hermite, in 1901; James Sylvester, in 1897; Arthur Cayley, in 1895; George Salmon, in 1904; Sophus Lie, in 1899; Francesco Brioschi, in 1897; Luigi Cremona, in 1903; Erwin Christoffel, in 1900; and Lazarus Fuchs, in 1902.

C. G. J. Jacobi.

Carl Gustav Jacob Jacobi (1804–1851). (Courtesy of the Universitätsbibliothek Heidelberg.)

But, the high point in the memory of the past was the commemoration of Jacobi. Indeed, the first lecture of the congress was *"Gedächtnisrede auf C. G. J. Jacobi,"* a biographical sketch of Jacobi by Leo Königsberger from Heidelberg. This lecture was printed and given to the participants as an official commemorating gift. Königsberger also wrote an extensive scientific biography of Jacobi, one that congress participants could purchase (due to a special offer from B. G. Teubner) at a third of its sale price. At the same ceremony, Schwarz gave an account of an episode regarding Jacobi's grave: some years before, the Prussian Academy of Sciences had been informed that Jacobi's elderly widow was not able to care for the grave of her husband, at Trinity

Church in Berlin. The Academy then took the necessary actions to buy the property and care for the grave. A cross and a plain but dignified fence were built. Schwarz presented to the congress a photograph of the current state of Jacobi's grave, giving it to Königsberger, as the official biographer, with the request that he keep it at his home. Four relatives of Jacobi were specially invited to this ceremony.

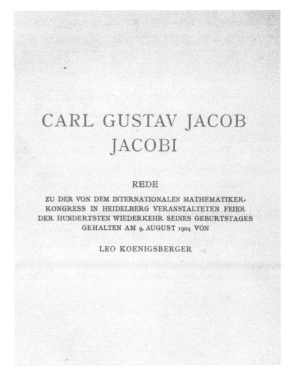

The 1904 congress commemorated the centennial of Jacobi's birth with a biography, which was offered to the members of the congress by the editor B. G. Teubner at a third of its sale price. (Courtesy of the Universitätsbibliothek Heidelberg.)

The scientific program followed the lines of the Zurich program. The organizing committee decided on four plenary lectures, one in each of the main languages of mathematics: German, English, French, and Italian. The lecturers were chosen by Weber, and their lectures were the following:

- "The Mathematical Theory of the Top (Considered Historically)," by Alfred George Greenhill from London;

- *"Le problème moderne de l'intégration des équations différentielles,"* by Paul Painlevé from Paris;

- *"La geometria d'oggidi e i suoi legami coll'analisi,"* by Corrado Segre from Turin;

- *"Riemanns Vorlesungen über die hypergeometrische Reihe und ihre Bedeutung,"* by Wilhelm Wirtinger from Vienna.

Originally, the French-speaking lecturer was to be Gaston Darboux, but after initially accepting, he later canceled his appearance.

The scientific sections were more or less similar to those of the previous congress:

- Section I: Arithmetic and Algebra,

- Section II: Analysis,

RÉCEPTION DE LINDBERGH   *Phot. N. Y. T.*

Paul Painlevé (1863–1933), plenary lecturer in 1904. (Courtesy of *The New York Times*.)

Alfred Greenhill (1847–1927), plenary lecturer in 1904. (Courtesy of the London Mathematical Society.)

Corrado Segre (1863–1924), plenary lecturer in 1904. (Courtesy of the Dipartimento di Matematica dell' Università degli Studi di Torino: *Fonti iconografiche* della Biblioteca Matematica "Giuseppe Peano.")

Wilhelm Wirtinger (1865–1945). (Courtesy of the Archiv der Universität Wien, Inv. 106.I.387.)

The congress witnessed very lively participation in the discussions taking place after each lecture. These discussions led to a changing of the regulations of the sections. At first, participants were allowed to take part only once in a discussion (and for no more than five minutes); now they were allowed to contribute twice (this regulation even applied to the lecturer!).

There is a famous incident related to one of these discussions. For the morning session of Wednesday, August 10, a lecture was scheduled entitled

- Section III: Geometry,

- Section IV: Applied Mathematics,

- Section V: History of Mathematics,

- Section VI: Pedagogy.

The only change was that of Section IV, which was devoted to Mechanics and Mathematical Physics in the previous congresses.

The total number of non-plenary lectures was very large: 78, more than double those of the previous congresses, although 13 mathematicians presented reports in more than one section. One of these was Hilbert, who presented *"Über die Grundlagen der Logik und der Arithmetik"* in the Arithmetic and Algebra section and *"Über die Anwendung der Integralgleichungen auf ein Problem der Funktiontheorie"* in the Analysis section. This large number of lectures forced the lectures in sections to be shortened from 30 to 20 minutes. As to the number of lectures in the sections, there was a balance; each section had roughly the same number of lectures.

Zum Kontinuum-Problem.*)

Von

J. König aus Budapest.

1. Es sei M_1, M_2, M_3, $\cdots$ eine abzählbar unendliche Folge beliebiger Mengen, deren Mächtigkeit wir mit $\mathfrak{m}_1$, $\mathfrak{m}_2$, $\mathfrak{m}_3$, $\ldots$ bezeichnen.

Mit Hilfe dieser Mengenfolge definieren wir zwei neue Mengen.

Die Summe der abzählbar unendlichen Mengenfolge, in symbolischer Bezeichnung:

$$S = M_1 + M_2 + M_3 + \cdots$$

bedeute jene Menge, die durch Zusammenfassung aller Elemente von M_1, M_2, M_3, $\cdots$ entsteht, wobei die verschiedenen Mengen angehörigen Elemente immer als voneinander verschieden anzusehen sind. Die Mächtigkeit von S bezeichnen wir mit $\mathfrak{s}$.

Das Produkt der abzählbar unendlichen Mengenfolge, in symbolischer Bezeichnung:

$$P = M_1 M_2 M_3 \cdots$$

bedeute jene Mengen, deren Elemente alle Komplexe

$$\mu = (\alpha_1, \alpha_2, \alpha_3, \cdots)$$

sind, wo α_i ein beliebiges Element der Menge M_i sein kann; es enthält demnach jedes μ ein und nur ein Element jeder beliebigen Menge der Folge. Es wird bequem sein, α_i als i-ten Index des Elementes μ zu bezeichnen. Die Mächtigkeit von P sei $\mathfrak{p}$.

Sind insbesondere alle M_i identisch $= M$, so wird statt P in der gebräuchlichen Bezeichnung $M^{\aleph_0}$ geschrieben.

Wir beweisen, daß, wenn die Mengen $M_1, \cdots$ transfinit sind**), immer die Beziehung

---

*) Für die hier benutzten Begriffe und Sätze sind die Arbeiten Georg Cantors, des Schöpfers der Mengenlehre, einzusehen. Insbesondere: „Beiträge zur Begründung der transfiniten Mengenlehre, I. und II" (Math. Annalen, Bd. 46 und 49).

Vgl. ferner A. Schoenflies: „Die Entwicklung der Lehre von den Punktmannigfaltigkeiten" (Jahresber. d. Deutschen Math.-Ver. VIII. 2).

**) Der Satz ist allgemeiner. Es besteht (1) dann und nur dann, wenn $\mathfrak{p}$

Jules König's lecture "On the Continuum Problem" was followed by a discussion with Cantor, Hilbert, and Schönflies, ICM 1904. (From the proceedings of the 1904 ICM, Teubner 1905.)

"Zum Kontinuum-Problem" (On the Continuum Problem) by the Hungarian mathematician Julius König from Budapest. The title directly addressed the first problem on the list presented by Hilbert at the 1900 Paris congress, "Cantor's Problem on the Cardinal Number of the Continuum." Word had spread through the congress, and interest was so high that the other sections canceled their meetings so that all participants could attend the lecture. König, in a polished lecture, concluded that Cantor's hypothesis was false. Both Cantor and Hilbert were present at the lecture. In the report of the proceedings called *Bericht über die Tätigkeit der Sektionen* (Report on the Activity of the Sections), we read that after the lecture there was a discussion in which Cantor, Hilbert, and Schönflies participated.

There is a famous recollection of those dramatic moments:

> Then Cantor spoke in a state of profound excitement. He expressed his gratitude to God's allowing him to live and see this refutation of his errors. The newspapers wrote of König's talk. The Archduke of Baden himself was informed of these sensational events by Felix Klein.

However, Cantor's impression was short-lived; the next day, Ernst Zermelo found the error in König's argument (which was due to a lemma by Bernstein).

Unhappily, this beautiful and moving story, reported by G. Kowalewski, has been questioned, particularly as to the discovery of the error as recently as the next day. (It seems that the discovery occurred a month later, and that Felix Hausdorff was involved). What is confirmed is that after the lecture, there was a tense informal meeting of several mathematicians in which the lecture was discussed in detail. In any case, we can still imagine the scene when Felix Klein, called in by the Archduke, had to explain the incident.

As a complement to the plenary lectures and the reports in the sections, the congress organized an "Exhibition of Literature and Models," with mathematical books, models, and apparatus that concentrated on recent materials, less than ten years old. The bibliographic exhibition consisted of scientific literature. More than 900 publications were exhibited with the help of publishers and individuals from many different countries; of these, 23 were German and 36 foreign. More then 300 mathematical models were shown, including cardboard polyhedra, wire mobile surfaces, wooden models, plaster models showing vibrating strings and heat diffusion, thread models for surfaces, and kinematic models for curves.

There were also mathematical instruments and apparatus. In this group, several individual exhibitors, academic institutions, and commercial companies presented different materials, some of them in the Applied Mathematics section. This made for a very lively session. The following devices were presented:

- the calculator "Triumphator" by Bombicki and Lamm from Berlin;
- the "Campylograph" and an angle divider by Chateau Frères, a mechanical precision company from Paris;
- the Coradi differentiator, the Abakanowicz integrator, the Payne–Coradi "Parabolograph," and a harmonic analyzer by G. Coradi from the Mathematic-Mechanical Institute of Zurich;
- a gyroscope (following Maxwell) and a gyrostat (following Lord Kelvin) by the Mathematical Institute of the University of Göttingen;
- the deformable Darboux hyperboloid by A. Greenhill from London;
- the "Brunsviga" and the "Addograph" calculators by Grimme, Natalis & Co. from Braunschweig;
- a "Cyclograph" and an "Ellipsograph" by the Technical School of Vienna;

the "Epidiaskop" and the "Episkop" by the optical workshop of Carl Zeiss from Jena.

There were demonstrations of several types of projecting lanterns, which were seen as future important auxiliary devices in teaching mathematics. Hermann Minkowski's lecture, *"Zur Geometrie der Zahlen,"* made use of images projected with the "Epidiaskop." The exhibition included explanatory talks and demonstrations, and many congress participants volunteered for this event.

The only historical model present at the exhibition was the calculating machine designed and built by Leibniz in 1674. There were only two existing copies (as today). The government of Hannover lent the one shown at the congress (the other one was in Munich). Carl Runge from Hannover explained how it operated. Unfortunately, it had design problems related to the adding of tens (9999 plus one would render 9900), but Leibniz had a correction mechanism devised for this situation. However, Runge explained that in its essential aspects, it coincided with a (then) modern calculator designed by Thomas.

The session of Thursday, August 11, held at the Aula of the University, began with the presentation of books and book projects. Not in vain did five publishers attend the congress: from Leipzig, A. Ackermann-Teubner (who, as we will see again when dealing with the finances, was deeply involved in the congress), W. Crayen, and R. Quelle; from Heidelberg, G. Köster; and from Paris, A. Gauthier-Villars.

The first book was *History of the German Mathematical Society*, written by A. Gutzmer from Jena and published by Teubner. Heidelberg was an appropriate place for this presentation, because it was the city where the project of a German mathematical society, stimulated by Cantor, had first been presented to the public. Each participant in the congress received a copy of the book.

At the 1904 congress, Carl Runge gave a lecture on Leibniz's calculating machine, built in 1674. (Courtesy of the Science & Society Picture Library.)

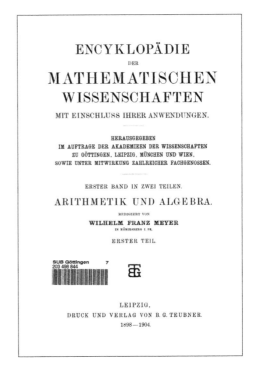

The first volume of the *Encyklopädie der Mathematischen Wissenschaften*. (Courtesy of the Niedersächsische Staats- und Universitätsbibliothek Göttingen.)

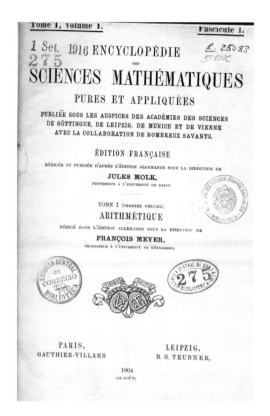

The first volume of the *Encyclopédie des sciences mathématiques*. (Courtesy of the Biblioteca de la Universidad Complutense de Madrid.)

Next, in the name of the responsible academic commission, Klein presented the first complete volume of the *Encyklopädie der Mathematischen Wissenschaften*, published by F. Meyer from Königsberg. The publication of this encyclopedia had begun in 1894. After that, Jules Molk from Nancy presented the first volume of the French edition of the encyclopedia. He explained that this was not a mere translation, but one that contained expository presentations on the articles of the German edition, written by French-speaking mathematicians. He praised the commitment of the editors, B. G. Teubner in Leipzig and Gauthier-Villars in Paris, and he explained that the motto chosen for the publication was *"Viribus unitis."*

The issue of the publication of Euler's works again attracted the attention of the International Congress. F. Morley from Baltimore and A. Vassilief proposed the following resolution:

> The III International Congress of Mathematicians, considering that the complete editions of the works of Euler has high scientific importance, supports the proposal presented to the Carnegie Institution by a mathematical committee presided over by M. Moore and hopes for its prompt resolution.

The assembly shared this interest and noted that steps in that direction were also being taken by the Academies of St. Petersburg and Berlin.

On Saturday, August 13, the last session of the congress began with the decision on the venue of the next ICM. Vito Volterra from Rome presented an invitation from the mathematical section of the Accademia dei Lincei and the Circolo Matematico di Palermo to celebrate the next congress in Rome in the spring of 1908 (this is the only ICM not held in the summer). The proposal was approved with enthusiastic applause.

Then, Alfred Greenhill, as now occurs when deciding the venue of the Olympic Games, stood and said:

> I left London with the impression that England was to be honoured with the visit of the International Congress of Mathematicians on the next occasion after Germany; ... Disappointed in our expectation we must congratulate Italy and Rome on its good fortune, and we must content ourselves with the next best in our wish, and hope that England may be selected at this Assembly as the meeting place in 1911 or 12.

Such a decision was not taken, since there was general agreement that the venue of a congress could only be decided at the closing of the previous one. The president of the congress, Weber, closed the congress with a recapitulating speech, ending with *"Auf Wiedersehn in Rom!"*

Nach Ländern geordnet ergibt diese Teilnehmerliste die folgende Gruppierung:

Land	Hauptkarten	Nebenkarten
Deutsches Reich	173	31
Rußland	30	4
Österreich-Ungarn	25	3
Frankreich	24	5
Vereinigte Staaten von Nordamerika	15	4
Dänemark	13	8
Italien	12	2
Schweiz	12	1
Schweden und Norwegen	8	1
Großbritannien	7	1
Niederlande	6	—
Belgien	2	—
Japan	2	—
Rumänien	2	—
Argentinien	1	—
Bulgarien	1	—
Canada	1	—
Griechenland	1	—
Spanien	1	—
	336	60

Im ganzen waren also 19 Länder durch 396 Personen vertreten.

Distribution of members of the 1904 Congress according to their nationality. (From the proceedings of the 1904 ICM, Teubner 1905.)

We have already commented on the increase of participation for this congress. As to the number of countries represented, the situation was different: they were 19. This is less than the 26 in Paris in 1900, but five more than in Zurich in 1897. The numbers of women among the participants also decreased; now there were only two: E. Maximova, *Gymnasiallehrerin* from Ustüchna, and M. Shilow, *Rechnerin an der Sternwarte* from Pulkowa.

The finances of the congress also reflect the nature of the congress: the government of Baden had granted a subsidy of 3000 marks; the Imperial Ministry of Medical, School, and Intellectual Matters (note this last responsibility of the ministry) gave 5000 marks; and His Majesty the Kaiser and King of Prussia contributed, from the funds at his free disposal, 5000 marks to attend Jacobi's commemoration. A particular donor was the company B. G. Teubner, "always willing to help our science," who provided 2000 marks. The participants were requested to pay a fee of 20 marks.

The congress was a success. H. W. Tyler, from the Massachusetts Institute of Technology, in his report to the *Bulletin of the American Mathematical Society*, praised the "efficient management of the Deutsche Mathematiker-Vereinigung … and the local arrangements, [which] both for the meetings and for hospitable welcome were generous and admirable."

ROME 1908

THE ROME CONGRESS witnessed several new features in the ICM series. The most relevant was the strong presence of applied mathematics and applications of mathematics. This presence was felt everywhere throughout the congress. The opening ceremony was held in the Sala degli Orazi e Curiazi of the Campidoglio, an ample hall magnificently decorated with frescoes by the Cavalier d'Arpino narrating the Roman legend of the fight between the Orazi and the Curiazi brothers. The ceremony was presided over by King Vittorio Emanuele III. There, in his speech, the Ministry of Public Instruction explained that:

The opening of the Rome 1908 congress (presided over by the King of Italy) took place in the Sala degli Orazi e Curiazi in the Campidoglio. (Courtesy of the Archivio Fotografico dei Musei Capitolini.)

The treatise of the European Economic Union was signed in 1957 in the Sala degli Orazi e Curiazi. (© European Community. Audiovisual Library of the European Commission.)

It is worth remembering the Italy of the urban republics and the Renaissance, with the names of Fibonacci, Tartaglia, del Ferro, Ferrari and so many others who prepared for the historical ripening of new spiritual and social demands for the blossoming of science. After Fibonacci, two streams appeared; one was seen in the studies of pure theory and the other grounded in the studies applied to commerce in which Italy was finding its renewed fortune. Thus, the double accounting of Luca Paciolo and his flourishing school of commercial arithmetic arose.

Applications of mathematics stem from the deep roots of Italian mathematics in the intense commercial activity of the late Middle Ages and also arose from the interest in problems arising in physics in the second half of the nineteenth century. These scientific origins matched up well with the interest of the Italian govern-ment. In addition to the collaboration of the Ministry of Public Instruction, the congress also obtained backing from the Ministries of Agriculture, Industry, and Commerce, of Finances, and of Public Development, which had explicitly suggested that the congress should not only be a congress on pure mathematics but also on applied mathematics.

This aim of reaching beyond pure mathematics is also seen when one looks at the institutions that helped finance the congress; here we find four Italian insurance companies. Four national associations of actuaries also sent delegates to the congress.

The presence of applied mathematics is also seen in the topics of some of the plenary lectures:

* *"Le partage de l'énergie entre la matière pondérable et l'éther,"* by Hendrik Antoon Lorentz from Leiden;

- *"La théorie du mouvement de la lune: son histoire et son état actuel,"* by Simon Newcomb from Washington;
- *"Sur les trajectoires des corpuscules électrisés dans le champs d'un aimant élémentaire avec applications aux aurores boréales,"* by Carl Størmer from Christiania.

Hendrik Antoon Lorentz had received the Nobel Prize in Physics in 1902 for a mathematical theory of the electron, and Simon Newcomb was a renowned astronomer from the U.S.A.

Simon Newcomb (1835–1909), plenary lecturer in 1908. (Courtesy of the American Mathematical Society.)

Hendrik Antoon Lorentz (1853–1928), Nobel Prize in Physics 1902, plenary lecturer in 1908. (Courtesy of the Emilio Segrè Visual Archives, American Institute of Physics.)

In each of the previous congresses, there had been four plenary lectures; in this case, the number increased to ten. The other seven lectures were

- *"Les origines, les méthodes et les problèmes de la géométrie infinitésimale"*, by Gaston Darboux from Paris;
- *"Die Encyklopädie der mathematischen Wissenschaften,"* by Walther von Dyck from Munich;

- "On the Present Condition of Partial Differential Equations of the Second Order, as Regards Formal Integration," by Andrew R. Forsyth from Cambridge;
- *"La mathématique dans ses rapports avec la physique,"* by Émile Picard from Paris;

Émile Picard (1856–1941), plenary lecturer in 1908. (From *Ouvres*, E. Picard, Centre National de la Recherche Scientifique Editions 1978.)

Gaston Darboux (1842–1917), plenary lecturer in 1908. (From *The Mathematician Sophus Lie* by Arild Stubhaug, Springer 2002.)

- *"L'avenir des mathématiques,"* by Henri Poincaré from Paris;
- *"Sur la représentation arithmétique des fonctions analytiques générales d'une variable complexe,"* by Gösta Mittag-Leffler from Stockholm;
- *"La geometria non-archimedea,"* by Giuseppe Veronese from Padua.

The opening lecture was entitled *"Le matematiche in Italia nella seconda metà del secolo XIX,"* by Vito Volterra. In the congress, Poincaré's health was a general concern when it was announced that his lecture would be read by Darboux. It is worth noting that Hilbert was invited but could not attend, also because of health reasons; in this case, the late communication prevented finding a substitute. Similarly, Klein was invited but had declined the invitation because of his multiple duties; he was replaced by Dyck.

The section of the previous congress on Applied Mathematics was expanded with the explicit decision to also consider Actuarial Mathematics. The full list of sections reveals the importance given to applied mathematics:

- Section I: Arithmetic, Algebra, and Analysis,
- Section II: Geometry,
- Section III (a): Mechanics, Mathematical Physics, and Geodesy,
- Section III (b): Various Applications of Mathematics,
- Section IV: Philosophical, Historical, and Didactical issues.

Moreover, it was Section III that had the most presentations. Among them, for example, "Notes on Steering of Automobiles and of the Balancing of Ships," by G. H. Brian from Upper Bangor, Wales.

Pietro Blaserna (1836–1918), president of the congress and of the Reale Accademia dei Lincei. (Courtesy of the Museo di Fisica dell' Universitá "La Sapienza" di Roma.)

Another novelty of the Rome congress was that for the first time at an ICM, an international prize was awarded. The story of this first prize is as follows. The congress was organized by two scientific institutions. One was the Reale Accademia dei Lincei, a legendary institution devoted to all branches of science, which had had Galileo Galilei among its members. The Accademia took care of the main duties for organizing the congress; its president, the physicist Pietro Blaserna, was president of the congress. The lectures were delivered at the Palazzo Corsini in the quarters of the Accademia (see page 158). The other institution was the Circolo Matematico di Palermo, which, as we have already seen, was one of the first mathematical societies, founded in 1884 by Giovanni Guccia.

Giovanni Battista Guccia (1855–1914), founder of the Circolo Matematico di Palermo. (Courtesy of the Circolo Matematico di Palermo.)

The contribution of the Circolo to the congress was twofold. On one hand, the Circolo was to take care of all the printing needed for the congress (announcements and proceedings). The other contribution was Guccia's offer, as president of the Circolo, of 3000 lire and a gold medal for an international prize, the *Medaglia Guccia*, to be awarded for a memoir containing substantial improvements on the theory of algebraic curves. This prize was announced at the 1904 Heidelberg congress; the jury was then formed by Max Noether from Erlangen, Henri Poincaré from Paris, and Corrado Segre from Turin. None of the three memoirs submitted was considered deserving of the award, which was then given, according to the regulations of the prize, to the work of Francesco Severi from Padua. Unfortunately, this was the first and only occasion when the *Medaglia Guccia* was awarded (Guccia still attended the 1912 congress in Cambridge and died in 1914).

The Reale Accademia dei Lincei was one of the institutions organizing the 1908 congress. (Courtesy of the Accademia Nazionale dei Lincei.)

The Sicilian Trinacria, logo of the Circolo Matematico di Palermo, one of the institutions organizing the congress. (Courtesy of the Circolo Matematico di Palermo.)

Francesco Severi (1879–1961), who was awarded the first (and only) *Medaglia Guccia*, a prize created for a memoir on algebraic curves, at the 1908 congress. (Courtesy of the Dipartimento di Matematica dell' Università degli Studi di Torino: *Fonti iconografiche* della Biblioteca Matematica "Giuseppe Peano.")

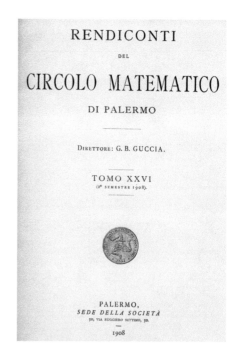

The *Rendiconti del Circolo Matematico di Palermo* planned to publish the proceedings of the 1908 congress, but a strike of Sicilian typographers prevented it from doing so. (Courtesy of the Biblioteca de la Universidad Complutense de Madrid.)

The other task of the Circolo, the printing of all the announcements, went well (2500 copies of the first and 4000 of the second). A problem arose, however, with the printing of the proceedings of the congress. The plan was to use the *Tipografia Matematica di Palermo*, where the scientific journal of the society, the *Rendiconti del Circolo Matematico di Palermo*, whose director was Guccia, was printed. However, a strike of the Sicilian typographers (the guild of typographers has been traditionally very combative, especially in Sicily and in the beginnings of the twentieth century) prevented the printing. The proceedings were finally printed in the *Tipografia dei Lincei*.

Spirits were high at the congress, and there were plenty of resolutions and proposals. At the suggestion of Jacques Hadamard, an international commission was created to study the unification of vectorial notation. Hadamard also suggested the possibility of celebrating together the international congresses of mathematics and physics. It was also proposed that the next congress study the creation of an international association of mathematicians. The issue of the publication of Euler's works was debated once more. There was even a proposal to create an archive of mathematical sciences.

But the most important and lasting of all these resolutions was the one stemming from the works of Section IV. Reports were presented in the sessions of the section considering the teaching of mathematics in secondary schools in many countries: Germany, France, England, Austria, the U.S.A., Hungary, Greece, Italy, and Spain. An idea conceived by David Eugene Smith from New York was then discussed and presented to the general assembly of the congress, where it was approved with lively applause:

> The Congress, recognizing the importance of a thorough examination of the programs and of the methods of teaching mathematics at secondary schools of different nations, charges Professors Klein, Greenhill, and Fehr to constitute an International Commission to study these questions and to report to the next Congress.

Thus, the origins of the Commission Internationale de L'Enseignement Mathématique, also known by its English name International Commission on Mathematical Instruction (briefly, ICMI), lay in the attention paid to educational issues at the Rome congress. The commission was linked from its beginning to the journal *L'Enseignement Mathématique*, founded in 1899, which since then has been its official journal.

In addition to the other developments at the Rome congress, there was an increase in size—both an increase in the number of participants and in the

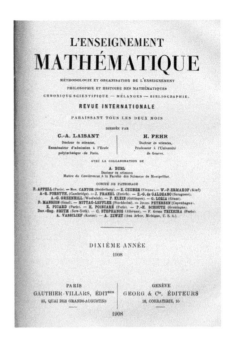

For many years *L'Enseignement Mathématique* has been the official journal of the International Commission on Mathematical Instruction. (Courtesy of the Biblioteca de la Universidad Complutense de Madrid.)

number of lectures. Total attendance was 535 participants, and 137 lectures were delivered, of which 10 were plenary (compare these figures to those of the Heidelberg cogress: 336, 78, and 4, respectively). The congresses were beginning to change from a restrictive reunion to a larger and more open meeting. This increase meant that the reports of the proceedings were more than 1000 pages, bound in three volumes. On the organizational side, the regulations created a very thorough procedure for creating the daily journal (which had also existed in the 1904 Heidelberg congress with the name *Tageblatt*). It was now mandatory to publish on the next day the titles of the papers read in the sections on the previous day; furthermore, the speakers were required to furnish the Secretary of the Section with a brief résumé of their lecture and the comments, immediately following the conclusion of the discussion.

Women were still participating but in very low numbers; ten were present, none of whom had attended previous congresses. There are two special cases, not included in the previous number. One is Emmy Noether, who was 26 years old and came accompanying her father, who was on the committee for the *Medaglia Guccia*. The other is the sad case of Dr. Laura Pisati. She was to present the communication "Essay on a Synthetic Theory for Complex Variable Functions," which would have been the first lecture given by a woman at an ICM. Unfortunately, eight days earlier, she died. In Section I, there was a memorial for her.

Distribuzione dei Congressisti per Nazioni.

	CONGRESSISTI	PERSONE DI FAMIGLIA	TOTALE
Italia	190	23	213
Germania	120	54	174
Francia	63	29	92
Austria-Ungheria	51	23	74
Inghilterra	22	11	33
Stati Uniti (America)	16	11	27
Russia	19	6	25
Svizzera	16	2	18
Svezia	5	1	6
Romania	6	—	6
Spagna	5	—	5
Danimarca	3	2	5
Olanda	3	1	4
Norvegia	2	2	4
Belgio	4	—	4
Grecia	3	—	3
Tunisia	2	—	2
Bulgaria	1	—	1
Canadà	1	—	1
Egitto	1	—	1
Messico	1	—	1
Serbia	1	—	1
TOTALE	535	165	700

Distribution of members of the 1908 congress according to their nationality. (From the proceedings of the 1908 ICM, R. Accademia dei Lincei 1909.)

The main financial contributor to the congress was the Italian government, with 11,600 lire. We have already noted that there was a contribution from insurance companies, a total of 1000 lire. Also, three publishers contributed a total of 400 lire. The amount left from the Heidelberg congress, almost 1000 lire, was incorporated into the budget of the congress. This reveals the deep feeling of continuity that the previous congresses had created.

Taking place in Rome, the congress could not avoid visiting renowned cultural sites. The city of Rome offered a reception in the Musei Capitolini in the Piazza del Campidoglio (refreshments were served, and "the halls were not cleared until midnight"). The congress also visited the Palatine, with its Roman sites. A buffet was served on the hill.

Aula Magna, Università "La Sapienza" di Roma. (Courtesy of the Dipartimento di Matematica "Istituto Guido Castelnuovo" dell' Università "La Sapienza" di Roma.)

The congress ended with the invitation by Andrew R. Forsyth to hold the next congress in Cambridge in 1912. The invitation was in the name of the Cambridge Philosophical Society, supported by the London Mathematical Society and many other English, Scottish, and Irish mathematicians. (There is no mention of the opinion of Welsh mathematicians.) The invitation was accepted with applause.

CAMBRIDGE 1912

It is true that there have been in the past of Cambridge great pure mathematicians such as Cayley and Sylvester, but we surely may claim without undue boasting that our University has played a conspicuous part in the advance of applied mathematics. Newton was a glory to all mankind, yet we Cambridge men are proud that fate ordained that he should have been Lucasian Professor here. But as regards the part played by Cambridge I refer rather to the men of the last hundred years, such as Airy, Adams, Maxwell, Stokes, Kelvin and other lesser lights, who have marked out the lines of research in applied mathematics as studied in this University.

THE INCLINATION toward applied mathematics of the Cambridge congress was perfectly expressed by Sir George H. Darwin (see page 48), president of the Cambridge Philosophical Society and president of

King's College, University of Cambridge. (Courtesy of Dr. Allan Doig and King's College. From *The Architectural Drawings Collection of King's College, Cambridge*, Avebury 1979.)

the congress (and son of Charles Darwin, the author of *On the Origin of the Species* and creator of the theory of evolution), at the opening meeting held on Thursday, August 22, at ten in the morning in the Examination Hall. As had occurred in the Rome congress with the Italian mathematical tradition, the Cambridge congress of 1912 was marked by the strong Cambridge and British tradition in applied mathematics.

Knowing that he was speaking to an audience composed almost entirely of pure mathematicians, G. H. Darwin ended his speech with the plea:

> I appeal then for mercy for the applied mathematician and would ask you to consider in a kindly spirit the difficulties under which he labours. If our methods are often wanting in elegance and do but little to satisfy the aesthetic sense of which I spoke before, yet they are honest attempts to unravel the secrets of the universe in which we live.

George H. Darwin (1845–1912), astronomer, president of the 1912 congress (son of the author of *On the Origin of the Species*). (Courtesy of The Royal Society of London.)

The influence of the applications of mathematics is again seen when one looks at the congress officials.

Indeed, the president of the congress was a renowned astronomer, who had presided over the Royal Astronomical Society, and the honorary president was John W. Strutt, better known as Lord Rayleigh, a renowned physicist, who had been awarded the Nobel Prize in Physics in 1904 for his discovery of argon gas.

John W. Strutt, Lord Rayleigh (1842–1919), honorary president of the 1912 congress and winner of the Nobel Prize in Physics in 1904. (Courtesy of The Royal Society of London.)

The opening of a mathematical congress in Cambridge could not end in those days without mentioning the famous Cambridge examination, the Mathematical Tripos. It was the Vice-Chancellor of the University who explained to the congress that:

> The Mathematical Tripos represented something like the oldest example in Europe of a competitive Examination with an order of merit . . . the Examination and the preparation for it has had a profound influence . . . on the study and progress of mathematics both in Cambridge and Great Britain.

The topics of the plenary lectures also show the influence of the applications of mathematics. There were eight plenary lectures, four of them devoted to pure mathematics:

"Boundary Problems in One Dimension," by Maxime Bôcher from Cambridge, Massachusetts;

"Définition et domaine d'existence des fonctions monogènes uniformes," by Émile Borel from Paris;

"Il significato della critica dei principii nello sviluppo delle matematiche," by Federigo Enriques from Bologna;

"Gelöste und ungelöste Probleme aus der Theorie der Primzahlverteilung und der Riemannschen Zetafunktion," by Edmund Landau from Göttingen.

The other four plenary lectures were devoted to applications of mathematics:

"Periodicities in the Solar System," by Ernest W. Brown from New Haven;

"The Principles of Instrumental Seismology," by Prince B. Galitzin from St. Petersburg;

"The Dynamics of Radiation," by Sir Joseph Larmor from Cambridge;

"The Place of Mathematics in Engineering Practice," by Sir W. H. White.

The influence of the applications of mathematics can also be seen in the list of sections of the congress. Compared to the previous congress in Rome, Geodesy was change to Astronomy, and Section III (b), before simply labeled Various Applications of Mathematics, was now very precisely specified to deal with Economics, Actuarial Science, and Statistics.

The 1912 congress visited the Cambridge Scientific Instrument Company. (Courtesy of the Syndics of Cambridge University Library, C.S.I.Co. Neg. 8973.)

The prime example of the strong inclination towards applications of the Cambridge congress was the visit to the Cambridge Scientific Instrument Company (see page 49). On the afternoon of Monday, August 26, after a full day devoted to sectional meetings and before being "entertained in the Hall and Cloisters of Trinity College by the Master and Fellows of the College," members of the congress visited the University Observatory and then went on to the works of the company. The company had been founded in 1881 by the young engineer Horace Darwin, the ninth son of Charles Darwin and brother of George H. Darwin. It was devoted to manufacturing high-quality scientific instruments. (The company had a long and productive life until 1968, when it was absorbed by a larger company in the same field of industrial activity.) The next day, the visit was repeated for other congress members. This trip is an illustration of the British scientific viewpoint; a thoroughly practical activity such as this was considered relevant to mathematicians.

Arthur Cayley (1821–1895). (Courtesy of the Staatliche Museen zu Berlin.)

This time, the traditional tribute to past figures in mathematics had a special character. The congress opened with the shadow of the recent death (a month before, on July 17) of Henri Poincaré. G. H. Darwin recalled the awarding to Poincaré the medal of the Royal Astronomical Society and cited his work *Science et Méthode,* referring to *"le sentiment de la beauté mathématique."*

Members of the 1912 congress visited Cayley's grave in Mill Road Cemetery, where they laid a wreath. (Courtesy of Wayne Boucher.)

Also very moving was another tribute. On the afternoon of Tuesday, August 27, after the lecture of Larmor, "a number of members of the Congress proceeded to the Mill Road Cemetery for the purpose of depositing a wreath upon the grave of the late Professor A. Cayley. An address was delivered by Professor Dickstein." As G. H. Darwin noted, this act touched the hearts of the University. It was decided that a silver wreath would be made and deposited in an appropriate place as a permanent memorial. The remaining money from the subscription of the laurel wreath and the white flowers was given to the organizing committee, which was entrusted with carrying out the project. Unfortunately, no trace of this wreath has been found in Cambridge.

As to publications of mathematical works, the congress expressed its warmest thanks to the Schweizerische Naturforschende Gesellschaft, the Swiss Society of Natural Sciences, for inaugurating the great work of the publication of the collected works of Euler in "magnificent style." This project had been backed by the whole mathematical community. It was first outlined in Zurich in 1897 by Rudio; the 1904 Heidelberg and 1908 Rome congresses supported the project. The Deutsche Mathematiker-Vereinigung contributed a third of its funds to this project, and the Académie des Sciences of Paris purchased 40 volumes (on the condition that works should appear in the original language). The Euler Commission in charge of the publication estimated that it would consist of forty volumes. The publication still continues; today, more than 70 volumes have been published. The Commission is also undertaking an edition of Euler's correspondence, which is expected to consist of five more volumes!

It was communicated to the congress that the Royal Society was in the process of publishing "a complete edition of the works of the immortal Herschel," the German-British astronomer from the eighteenth century, responsible for the discovery of Uranus. In the History Section, it was announced that an edition of the collected works of the historian of mathematics Paul Tannery was in preparation by Sir Thomas L. Heath, who had just published his renowned canonical edition of Euclid's *Elements*.

As was becoming a tradition at the ICM, the congress organized an "exhibition of books, models and machines (chiefly calculating machines)," which was "arranged in two rooms of the Cavendish Laboratory" (see page 52).

One of the duties of the congress was to review the state of the resolutions approved at the Rome congress. The congress expressed its appreciation for the impressive work done by the International Commission on the Teaching of Mathematics, led by its oddly entitled Central Committee (formed by F. Klein, A. Greenhill, and H. Fehr) with the aid of D. E. Smith. They reported that "[e]very country in nearly every part of the world has contributed in its own department to the *Reports for Cambridge*—so that there were about 150 different volumes with about 300 articles brought to the Congress."

Another proposal of the Rome congress was to establish an International Association of Mathematicians. It was the president of the congress who addressed this issue and, in line with the British traditions, said, "Our existing arrangements for periodical congresses meet the requirements of the case better than would a permanent organization of the kind suggested." Regard-

The first volume of the collected works of Euler published in 1911. (Courtesy of the Biblioteca de la Universidad Complutense de Madrid.)

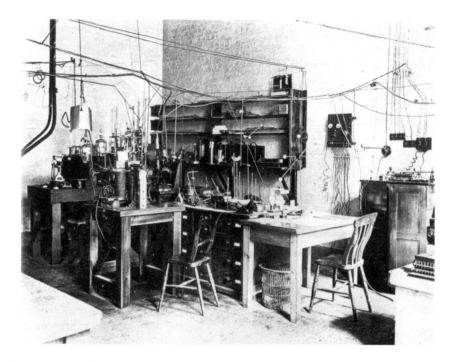

The Cavendish Laboratory. For the 1912 congress, it hosted an exhibition of "books, models and machines (chiefly calculating machines)." (Courtesy of the Emilio Segrè Visual Archives, American Institute of Physics.)

less of this final outcome, it is quite astonishing that when the political and social atmosphere of Europe was progressively worsening (two years later the war would break out), mathematicians could seriously consider this issue. However, as to the Rome proposals, there was a failure to be reported: in regard to the intention of unifying the vectorial notation, Hadamard sadly reported on the impossibility of arriving at any agreement. If one had to choose an issue on which to disagree, this was probably one of the most innocent choices.

The Cambridge congress represents the highest point for the ICM since its beginnings. Participation reached 574 members coming from 28 countries. Although European presence was dominant, there were 82 members coming from nine non-European countries. Mathematicians from the U.S.A. were the sec-

ond largest national group after the British with 60 members. The language issue was carefully dealt with, and the regulations of the congress were written in English, French, German, and Italian. The Leipzig editor A. Ackermann-Teubner, a fixture at all of the previous congresses, also attended.

This congress was a landmark for the participation of women. Section II on Geometry witnessed in its session of Monday, August 26, the first communication presented by a woman; it was "On Binodes and Nodal Curves," by Miss H. P. Hudson. The participation of women was much higher than in other congresses: there were 38 women, 30 from the United Kingdom. The other eight were Professor Mademoiselle Anna Amieux from Paris, accompanied by her mother; Mademoiselle Byck from Kiev, accompanied by two family members; Madame Marie Cher from

	Members	Members of family	Total
Argentine	5	—	5
Austria	20	3	23
Belgium	5	—	5
Brazil	1	—	1
Bulgaria	1	—	1
Canada	5	—	5
Chili	1	—	1
Denmark	4	2	6
Egypt	2	—	2
France	39	6	45
Germany	53	17	70
Greece	4	1	5
Holland	9	1	10
Hungary	16	3	19
India	3	—	3
Italy	35	6	41
Japan	3	—	3
Mexico	2	—	2
Norway	3	1	4
Portugal	2	1	3
Roumania	4	1	5
Russia	30	10	40
Servia	1	—	1
Spain	25	2	27
Sweden	12	2	14
Switzerland	8	2	10
United Kingdom	221	49	270
United States of America	60	27	87
	574	134	708

Distribution of participants in the 1912 congress according to nationality. (From the proceedings of the 1912 ICM, Cambridge University Press 1913.)

Mittag-Leffler, in the name of the Swedish Academy and the journal *Acta Mathematica*, invited the 1912 congress to hold its next meeting in Stockholm in 1916. (From *The Mathematician Sophus Lie* by Arild Stubhaug, Springer 2002.)

Paris; Dr. Elizabeth Cowley from Poughkeepsie, New York; Mademoiselle Dr. Renée Masson from Geneva; Miss Marion Reilly from Bryn Mawr, Pennsylvania; and Professor Ruth G. Wood from Northampton, Massachusetts.

The final meeting of the congress was held on Tuesday, August 27, at nine in the evening. G. H. Darwin gave one of the most original speeches thanking the organizing committee. He noted the historical context of the meeting: "You are perhaps aware that our Parliament in its wisdom has decided that coal-miners shall not be allowed to work for more than eight hours a day. There has been no eight hours bill for the Secretaries of this congress."

After him, following the regulations for the congresses and the established tradition, Mittag-Leffler presented the invitation for the next congress:

In the name of the first class members of the Royal Academy of Sciences, in the name of the Swede redaction of the journal *Acta Mathematica* and in the name of the Swede geometers, I have the honor of inviting the International Congress of Mathematicians to meet in Stockholm in 1916.

This was the same invitation he had already presented four years before in 1908 in Rome. There were also two other invitations extended at that time: E. Beke, in the name of "the homeland of Bolyai," offered to host the seventh congress in Budapest in 1920, and C. Stéphanos expressed the hope that the congress would meet in Athens in 1920 or 1924.

For many participants, the congress was a purely British experience: accommodation was offered at the

Acta Mathematica. (Courtesy of the Biblioteca de la Universidad de Sevilla.)

colleges of the university; Newham College was reserved for ladies. Different receptions were held: in Hatfield House at the invitation of the Marquis of Salisbury, and for *conversazione* in Fitzwilliam Museum at the invitation of the Chancellor of the University, Lord Rayleigh. And it was reported that it rained almost every day.

A very evocative briefing was written by J. W. A. Young from Chicago for *The American Mathematical Monthly*:

> Meeting in a University which for rare beauty and charm of picturesque medieval buildings and exquisite gardens, can find a rival only in its sister university of Oxford; living in the rooms occupied in times past by generation after generation of the world's greatest savants, dining in halls of storied interest from whose crowed walls look down the likenesses of great men of many ages whom the world still delights to honor; royally entertained with that cordial hospitality for which Englishmen are so justly famed, the mathematicians who gathered at Cambridge from the four corners of the world lived through a unique week of their lives, and carried away with them a souvenir, never to be forgotten, of a congress brilliant alike in historic and lovely surroundings, in elaborate social functions, and in number and value of mathematical lectures, reports and papers presented.

At that moment, the future of the ICM as the wellspring of international cooperation in mathematics seemed sure and clear.

Interlude

IMAGES OF THE ICM

IF HILBERT'S LECTURE in the Paris 1900 congress is the mathematical icon of the International Congress of Mathematicians, the icon of graphics is the illustration used as the frontispiece of the proceedings of the Zurich 1897 congress (which also served as an identification card for the congress participants).

The magnificent graphic composition is a lithography with the images of five great Swiss mathematicians. Presiding over the five is Jacob Bernoulli (Basel, 1654–1705), patriarch of the mathematical family of the same name and author of the first classical treatise on probability theory, *Ars Conjectandi*. To his left is his brother Johann (Basel, 1667–1748), who was a staunch defender of Leibniz's claims of priority in the discovery of calculus against the claims of Newton and his followers (the historian of mathematics Carl B. Boyer calls him "Leibniz's bulldog"); Johann Bernoulli's severe image is that of the person who was engaged throughout his life in continuous and innumerable mathematical controversies (even with his brother and his son). To Jacob's right is Daniel Bernoulli (Basel, 1700–1784), son of Johann and developer of the applications of the Leibnizian calculus to the study of many mechanical problems (among them, that of the vibrating string). Below Daniel is Leonhard Euler (Basel 1707, St. Petersburg 1783), possibly the greatest eighteenth-century mathematician, whose prolific production we have already referred to. He is the author of the

Lithography in the proceedings of the 1897 Zurich congress featuring five great Swiss mathematicians. (From the proceedings of the 1897 ICM, Teubner 1898.)

classical treatise *Introductio in analysin infinitorum*. Opposite Euler is Jakob Steiner (Utzenstorf 1796, Bern 1863), author of the influential treatise on projective geometry, *Systematische Entwicklungen*, and known as "the greatest geometrician since the time of Apollonius."

This mathematical scene is complemented by a drawing of the facade of the Eidgenössische Polytechnikum (now called Eidgenössische Technische Hochschule or ETH), venue of the Zurich congress, and for decades heart of the international collaboration in mathematics (three international congresses have been held in the ETH: those of 1897, 1932, and 1994; and three presidents of the International Mathematical Union came from the ETH: Heinz Hopf, Komaravolu Chandrasekharan, and Jürgen Moser). The illustration is ornately decorated with a variety of typical Swiss flowers, notably the edelweiss from the Alps.

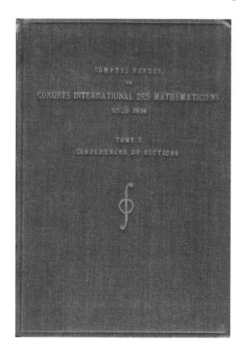

The "Oslo integral," ICM 1936. (From the proceedings of the 1936 ICM, A.W. Brogers Boktrykkeri A/S 1937.)

The Zurich illustration set a high standard for the graphic creations associated with the international congresses. The next appearance of an ICM graphical creation had to wait until the Oslo 1936 congress. The trail is tenuous; the graphic is a simple complex circle integral printed on the front cover of the proceedings volume. However, the use of the design was extensive; it was reported by G. Waldo Dunnington (Gauss' famous biographer) that "delegates to the Congress wore a badge in the form of an integral sign, which entitled them to ride free on street cars and buses in Oslo and vicinity." (Here we find one more among the multiple applications of the integral!)

The "Moscow integral," ICM 1966. (Courtesy of the International Mathematical Union.)

This was the beginning of a series of international congresses with mathematically flavored graphic logotypes. The Moscow 1966 congress created a logotype based on an integral sign combined with the flat image of a sphere where parallels and meridians are drawn. This logo was used throughout the congress, in the participant's badge and in banners displayed at the opening and in a huge standing structure built outside the University of Moscow. This was the first time an international congress used such intense advertising-propaganda techniques. The logo was immortalized in a famous stamp issued at the time of the congress.

The Moscow stamp, ICM 1966.

The "Helsinki disk," ICM 1978. (Courtesy of the International Mathematical Union.)

The Helsinki 1978 congress chose as a logo a Poincaré disk colored in intense blue with parallel lines (for the Poincaré metric) highlighted. A different set of parallel lines was drawn on the disk for the version of the logo shown in the congress stamp. A card was sold at the congress with a beautiful combination of the different versions of the Poincaré disk appearing in the logotype, the stamp, and the postmark.

The Helsinki stamp, ICM 1978.

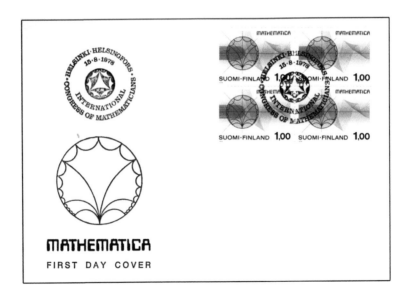

The Helsinki postmark, ICM 1978.

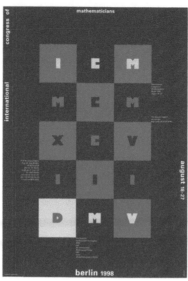

Berlin posters, ICM 1998. (Courtesy of the poster designers Ott & Stein.)

A combination of the Roman numerals for the number 1998 and the letters for the acronym ICM, designed by Ott & Stein, was the motif for the logo and poster for the Berlin 1998 congress. There was also another more colorful poster with colored surfaces.

The Berlin stamp, ICM 1998.

The stamp issued for this congress, full of mathematical substance, was very popular. It features the number 110, which can be represented as the sum of three squares exactly in three different ways (this is a good exercise; to be solved by the reader before the paragraph ends); the colored squares (note that only four colors are used) form a "near square" with sides 177 and 176 (in appropriate units). Lastly, the graphic artist who designed the stamp (Norbert Höchtlen) completed the drawing representing the decimal expansion of π forming a sequence displayed in concentric rings (not so easy to appreciate in the picture).

The Beijing 2002 congress created a beautiful logotype based on a diagram drawn by Zhao Shuang, a third-century Chinese mathematician from the Zhou Dynasty, demonstrating the Pythagorean theorem. Here is the official explanation of the "inspirations" needed to transform the diagram into the logo:

First, by opening the edge of the outer square and enlarging the square inside, it will symbolize that the minds of mathematicians are open, and that China is open. Next, varying colors make the diagram more like a rotating pinwheel to symbolize the hospitality of Beijing people. (The pinwheel is a folk toy which you may see children in Beijing's *hutong* playing with and greeting you: "Welcome, welcome!")

The Beijing logo, ICM 2002. (From the ICM 2002.)

(*Hutongs* are alleys in Beijing's old quarters.) This logo appeared in stamps, posters, and also on a congress postcard, which included a festival of numerals, as well as the ancient Chinese numeral system based on (bamboo) rods.

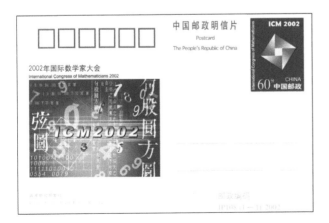

The Beijing postcard, ICM 2002. (From the ICM 2002.)

The sunflower symbolizes the Spain of sun and light … the number of its spirals to right and left are elements of the Fibonacci sequence … an image that resembles both a sunflower and the fractal nature of a Romanesco cauliflower.

This is not the transcription of the broken mumbling of a mathematician when hallucinating, but the official description of the logotype of the Madrid 2006 congress. It is a challenge (solvable?) to find all those elements in the image of the logo (see page 60).

The Madrid poster, ICM 2006. (Courtesy of the ICM 2006.)

The Madrid "Fibonacci sunflower," ICM 2006. (Courtesy of the ICM 2006.)

Amsterdam's Grachtenhuis, ICM 1954. (Courtesy of the Het Koninklijk Wiskundig Genootschap in Amsterdam.)

Logo of the Vancouver congress, ICM 1974. (Courtesy of the Canadian Mathematical Society, formerly the Canadian Mathematical Congress.)

Totem pole from the Haida people, from the western coast of Canada. (Courtesy of the University of British Columbia Archives.)

Other congresses have taken a totally different direction when designing their graphic images, choosing traditional motifs from the national cultures. The first example of this was the stylish logotype of the Amsterdam 1954 congress; it shows the shape of a Grachtenhuis, the seventeenth-century merchant's house that can be seen when one walks around Amsterdam's canals.

The impressive head of a totem from the Haida people (an ancient culture from the western coast of Canada) was the official congress insignia of the Vancouver 1974 ICM. In fact, the University of British Columbia, host of the congress, has an extraordinary Museum of Anthropology with many totem poles, the inspirational source for the logo.

The Kusudama, ICM 1990.

traditional use of the number 10^{16}? The commemorative stamp for the congress shows the *Kusudama*, a Japanese tradition of making three-dimensional bodies by folding paper sheets with different colors. The congress poster reflects the image of silence and mystery associated with the Far East.

The Kyoto "stone lantern," ICM 1990. (Courtesy of ICM 1990.)

The Kyoto 1990 congress iconography displayed several traditional Japanese motifs. The most recognizable was the congress logotype, present in all printed materials of the congress, designed by Kazuyoshi Aoki and Yuji Komai. It "symbolizes a Japanese stone lantern, the first letter for Kyoto, as well as the character for 10^{16}." This last statement defies rational thinking: if the symbol is of ancient origin, what was the

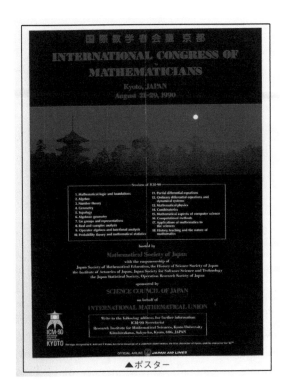

The Kyoto poster, ICM 1990. (Courtesy of ICM 1990.)

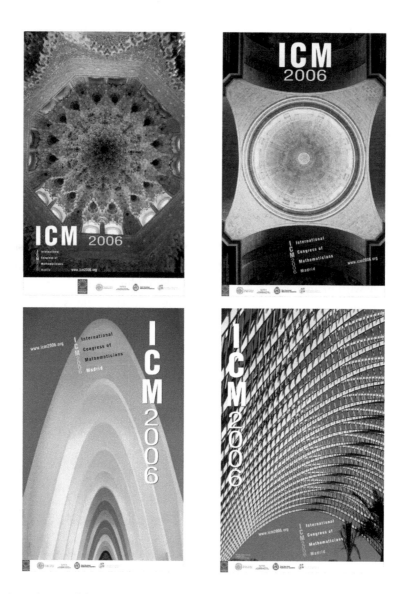

Madrid posters were based on well-known pieces of Spanish architecture with mathematical content: La Alhambra, El Escorial, Colegio de las Teresianas, Ciudad de las Artes y las Ciencias, ICM 2006. (Courtesy of the ICM 2006.)

The Madrid 2006 congress issued a series of posters showing well-known pieces of Spanish architecture with mathematical content: the Muslim Palace of the Alhambra in Granada, the Monastery of El Escorial near Madrid, the "Colegio de las Teresianas" by the ar-chitect Antoni Gaudí in Barcelona, and the "Ciudad de las Artes y las Ciencias" by the engineer Santiago Cala-trava in Valencia. The Madrid commemorative stamp featured the first known written record of the entire set of the Hindu-Arabic numerals (from the Medieval

Codex Vigilanus, which we will discuss in "Social Life at the ICM").

The Madrid commemorative stamp featured the first known written record of the entire set of the Hindu-Arabic numerals, ICM 2006.

Three congresses chose to have an image based on pure graphic design. The first was the Warsaw 1982 congress. In the ICM 1982 logotype, there are certain mathematical reminders, but the emphasis is placed more on the graphical design than on the possible mathematical interpretation. It was designed by Stefan Nagiełło.

The Warsaw logo, ICM 1982. (Courtesy of the Instytut Matematyczny Polskiej Akademii Nauk.)

In conjunction with the Warsaw congress, stamps featuring famous Polish mathematicians were issued.

Stamps featuring famous Polish mathematicians.

Similar to the Warsaw logotype, the one designed for the Berkeley 1986 congress might have some mathematical interpretation, but what stands out to the viewer is its commercial tone. It looks like an advertising image for a household product. The power of the strongly commercial U.S. economy is always present!

The Berkeley logo, ICM 1986. (Courtesy of the International Mathematical Union.)

The gem of graphic design is the logotype of the Zurich 1994 congress, designed by Georg Staehelin.

The Zurich logo, ICM 1994. (Courtesy of the Swiss Mathematical Society.)

It is a stylish combination of the name of the city and the acronym of the congress.

The poster of the congress is a collage-type composition of images of the city, the lake, and the Grossmünster church of Zurich.

The Zurich poster, ICM 1994. (Courtesy of the Swiss Mathematical Society.)

ICM acronym, ICM 1962, Stockholm. (Courtesy of the Archives of the International Mathematical Union at the University of Helsinki.)

What about the acronym ICM for the International Congress of Mathematicians? Its first appearance was in the Stockholm 1962 congress, where the official paper, program, and abstract booklets carried the acronym. But, as we saw before, it was the Moscow 1966 congress that popularized the acronym. The flyleaves of the proceedings were decorated with a beautiful Grecian fret composed of the letters ICM and the name Moscow written in different languages.

The ICM acronym in the proceedings of the Moscow 1966 congress. (From the proceedings of the 1966 ICM, Mir 1968.)

The logo of the next congress, which will be held in Hyderabad, India, in 2010, is already available. It is a beautiful blending of mathematics and tradition:

The logo for ICM 2010 depicts the standard fundamental domain for the modular group $SL(2, \mathbb{Z})$ acting on the upper half plane. The formula written along the circular arc is a famous conjecture of the Indian mathematician Srinivasa Ramanujan proved by Pierre Deligne in 1973. The quotation in Sanskrit at the bottom of the logo is from the *Rig Veda*, an ancient Indian religious work dating back to more than 1000 years

The logo of ICM 2010, to be held in Hyderabad, India. (Courtesy of ICM 2010.)

before the start of the Christian era. It translates as "May good ideas come to us from everywhere."

Let us end this first interlude with the new logo adopted by the International Mathematical Union in 2006 and presented at the Madrid 2006 congress. The logo design is based on three rings forming the so-called Borromean rings. They have been used over many centuries and in many cultures. The rings have the property that if any one of its components is removed, the other two can fall apart, while all three together remain linked. Although the Borromean rings are often drawn as if made from three round circles, such a construction is mathematically impossible. The designer, John Sullivan, from the Technical University of Berlin, explained the meaning of the new logo: "It represents the interconnectedness not only of the various fields of mathematics, but also of the mathematical community around the world."

The new IMU logo. (Courtesy of the International Mathematical Union.)

PART II
CRISIS IN THE INTERWAR PERIOD

O N WEDNESDAY EVENING, July 15, 1936, the City of Oslo is offering a dinner at the Bristol Hotel for the members of the International Congress, which has convened in the city. Several speeches are delivered, beginning with a representative from the municipality, who greets the guests. The organizing committee has prepared speeches in different languages. In the name of the German-speaking members of the congress, Erhard Schmidt from Berlin recalls the relation of the great Norwegian mathematicians Niels Henrik Abel and Sophus Lie with the German universities; for the English-speaking members of the congress, Luther P. Eisenhart from Princeton stresses that "mathematics is international . . . it does not recognize national boundaries"—an idea that, though clear to mathematicians throughout time, was subject to questioning at those moments. Lastly, the French mathematician Gaston Julia, professor at the Université de la Sorbonne in Paris, takes the stand as the French voice. After praising the country of Amundsen, Ibsen, and Grieg, he evokes a personal story:

GASTON J U L I A
en 1919

Gaston Julia (1893–1978), who was seriously injured in World War I, was forced to wear a mask over his face for the rest of his life. In his speech to the 1936 congress, he recalled those dramatic moments in the war. (From *Ouvres*, G. Julia, Gauthier-Villars 1968.)

Twenty years ago, a young, wounded officer was taken after having surgery at night to a room. He was falling asleep when he was awakened by his own blood overflowing in his mouth: an artery had just reopened. He barely had time to cry for help before losing consciousness.

When he recovered consciousness, he recognized the nurse in charge of the service by his side. In the absence of the surgeon, who had left the hospital, and of the night doctor occupied elsewhere at that moment, without hesitating she instantly stopped the bleeding with confidence and determination, reanimating that fainting body. When the doctor returned he realized everything was well done and praised the decision and ability of the nurse.

Because of the fear of the accident's occurring again, in a spontaneous and charitable gesture, that generous lady decided to remain that difficult night by the wounded soldier. I will never forget that long night in which, almost unable to speak, weakened by the bleeding, and unable to get sleep, I felt relieved by the presence of that woman who, sitting by my side, was sewing in silence under the discreet circle of light from the lamp, listening at regular intervals to my breathing, taking my pulse, and looking into my eyes, which only by glancing could express my ardent gratitude.

Ladies and Gentlemen, this generous woman, this strong woman was a daughter of Norway.

What could have happened to justify such a speech at a ceremony in a congress? Clearly it was World War I, also know as the Great War. Beyond the impressive intensity of the personal tribute contained in these words, the scene has a deep significance when interpreted within the history of the international congresses. It is a good point for appreciating the difficulties and problems that occurred for the international mathematical cooperation in the period between the two world wars.

In this period, the story of the international congresses cannot completely match the idyllic image reported in 1912 for the scientific journal *Science* by A. R. Crathorne from Illinois:

> Once every four years the mathematicians of the world meet together to discuss the new discoveries made in the various branches of their science, to review the work accomplished during the past quadrennial period, to listen to mathematical papers and to become acquainted with one another.

The pressure from the "outside world" was too strong and tainted the course of the congresses. The five congresses held between the two world wars were somewhat turbulent. They were

- Strasbourg, September 22–30, 1920;
- Toronto, August 11–16, 1924;
- Bologna, September 3–10, 1928;
- Zurich, September 5–12, 1932;
- Oslo, July 14–18, 1936.

STRASBOURG 1920

I agree with you that we as mathematicians need to be at the head in "the task of reestablishment of friendly relations" between the men of science of all countries.

THESE LINES ARE FROM A LETTER from Mittag-Leffler to Hardy on January 25, 1919. They reveal both noble intentions and a harsh situation. Unfortunately, the intentions turned out to be less powerful, and the situation harsher, than expected.

Stained-glass window from Strasbourg's Cathedral. (From *La cathédrale de Strasbourg,* Hannesschläger 1970.)

The Great War of 1914–1918 and its aftermath had a tremendous impact on all aspects of social life; science was not immune to it. In Brussels in 1919, the Allied Powers created the International Research Council (IRC, in short), whose objectives were

1. to coordinate international efforts in the different branches of science and its applications,

2. to initiate the formation of international associations or unions deemed to be useful to the progress of science.

Up to this point, the objectives look fairly reasonable, but a further specification referred to a resolution adopted in London in October 1918, before the war had ended: "without delay by the nations at war with the Central Powers, with the eventual co-operation of neutral nations."

A further detail reveals the true nature of the IRC: the members of the Council were not scientists or scientific associations, but nations, better to say governments, and, at least at the beginning, just those of the Allied Powers. The almost publicly declared objective of the Council was to eliminate the preeminence that German science had in many fields. Well-known mathematicians had a relevant role in the IRC: Émile Picard was President, until 1931, and Vito Volterra Vice-president.

Émile Picard (1856–1941), honorary president of the Union Mathématique Internationale, founded at the 1920 congress. (Drawing made at the congress by an eminent Strasbourg artist.) (From the proceedings of the 1920 ICM, Eduard Privat 1921.)

In the Brussels 1919 meeting, two important decisions were taken with regard to international collaboration in mathematics. The decision adopted in Cambridge in 1912 to celebrate the next international congress in Stockholm was overturned, and an option more along the lines of the Treaty of Versailles was taken: the congress would be held in Strasbourg, capital of the region of Alsace, just regained by France after its loss to Germany in the Franco-Prussian war of 1870–1871. Also, the draft of the statutes of the Union Mathématique Internationale (UMI, in short) was written. Members of the union would be countries represented via a mathematical national committee to be established.

Thus, following IRC instructions, in September 1920 a mathematical congress took place at the Université de Strasbourg. The president of the congress was Picard, and the honorary president the 82-year-old Camille Jordan (see page 74).

The congress was comprised of five plenary lectures:

- "Questions in Physical Indetermination," by Sir Joseph Larmor from Cambridge;
- "Relations between the Theory of Numbers and Other Branches of Mathematics," by Leonard Eugene Dickson from Chicago;

Leonard E. Dickson (1874–1954), plenary lecturer in 1920 and Vice President of the UMI. (Drawing made at the congress by an eminent Strasbourg artist.) (From the proceedings of the 1920 ICM, Eduard Privat 1921.)

- "Sur les fonctions à variation bornée et les questions qui s'y rattachent," by Charles de la Vallée Poussin from Leuven (see page 71);
- "Sur l'enseignement de la physique mathématique et de quelques points d'analyse," by Vito Volterra from Rome (see page 71);
- "Sur les équations aux différences finies," by Niels Erik Nörlund from Lund;

and 79 communications distributed in the four classical sections:

* Section I: Arithmetic, Algebra, and Analysis,
* Section II: Geometry,
* Section III: Mechanics, Mathematical Physics, and Applied Mathematics,
* Section IV: Philosophical, Historical, and Pedagogical Issues.

The congress was a long one, September 22 to 30, but the level of activity was moderate. Of the nine congress days, one full day was devoted to the opening session, and five days had sessions of the sections in the morning and a plenary lecture in the afternoon; activity was ended by 4 p.m. There were many receptions: the first congress day, at the events hall of the university;

Vito Volterra (1860–1940), plenary lecturer in 1920 and Honorary President of the UMI. (Drawing made at the congress by an eminent Strasbourg artist.) (From the proceedings of the 1920 ICM, Eduard Privat 1921.)

Charles de la Vallée Poussin (1866–1962), plenary lecturer in 1920 and President of the UMI. (Courtesy of Olli Lehto.)

At the 1920 congress, there was a lecture on the astronomical clock of Strasbourg's Cathedral (the clock was made in 1354). (From the proceedings of the 1920 ICM, Eduard Privat 1921.)

the next day, a tea at the Society of Friends of the University; one day later, another official reception (with tea) at City Hall; and one more reception the next day, this time hosted by the Commissary General of the Republic (also with tea). The feast ended with a farewell banquet at the Baeckehiesel Restaurant. And three full days were devoted to excursions!

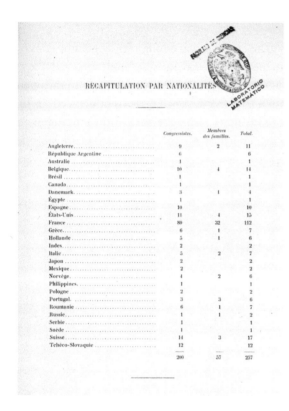

	Congressistes.	Membres des familles.	Total.
Angleterre	9	2	11
République Argentine	6		6
Australie	1		1
Belgique	10	4	14
Brésil	1		1
Canada	1		1
Danemark	3	1	4
Égypte	1		1
Espagne	10		10
États-Unis	11	4	15
France	80	32	112
Grèce	6	1	7
Hollande	5	1	6
Indes	2		2
Italie	5	2	7
Japon	2		2
Mexique	2		2
Norvège	4	2	6
Philippines	1		1
Pologne	2		2
Portugal	3	3	6
Roumanie	6	1	7
Russie	1	1	2
Serbie	1		1
Suède	1		1
Suisse	14	3	17
Tchéco-Slovaquie	12		12
	200	57	257

Distribution of participants in the 1920 congress according to nationality (observe the presence of Poland and Czechoslovakia, some of the new countries created after the Treaty of Vesailles). (From the proceedings of the 1920 ICM, Eduard Privat 1921.)

The number of participating countries was 27, just one less than in Cambridge in 1912. Reading the list of countries, we find some of the consequences of the Treaty of Versailles: there were new countries, Poland and Czechoslovakia. In order to judge the success of the

congress, the number of countries participating is misleading because the number of mathematicians attending was only 200 (and 80 were French). This makes this congress the one with the least number of participants in the history of the ICMs. What was the reason for this? Had the Great War crushed the enthusiasm of the prewar ICMs?

No, the reason for this low attendance was twofold. On one hand, it was the exclusion of mathematicians from the former Central Powers (Germany, the Austro-Hungarian Empire, Bulgaria, and Turkey) imposed by the IRC. The secretary general of the congress, Gabriel Koenigs from Paris (see page 73), explained the procedure for choosing the participants: "The Congress has been convened by not collective but individual invitations, sent by the own French national committee, who also centralized the proposals for lectures or communications."

VENDREDI 24 SEPTEMBRE

De 9 à 12 heures. — Séances ordinaires des Sections.
10 heures. — Visite du Mausolée du Maréchal de Saxe.
11 heures. — Visite de la Cathédrale. (Ces deux visites sous la conduite de membres de la Société des Sciences du Bas-Rhin.)
14 h. 1/2. — Conférence générale de M. Dickson (Salle des Fêtes).
17 h. 1/2. — Réception officielle à l'Hôtel-de-Ville (thé). Allocution de M. H. Lévy au nom de la ville de Strasbourg, de M. G. Koenigs et de M. C. de la Vallée-Poussin au nom des Congressistes. Mme Jeanne Clapier, de l'Académie de Vaucluse, récite une Ode « Salut à Strasbourg » de la plus belle inspiration.
20 h. 1/2. — Séance organisée en l'honneur du Congrès par la Société des Sciences du Bas-Rhin (Salle de l'Aubette, place Kléber). Programme : Conférence de M. le Général Taufflieb, Sénateur, sur : « La Science en Alsace »; Allocutions, Concert.

Repas. — Les Congressistes officiers de réserve dans une Armée alliée sont avisés qu'ils peuvent prendre leurs repas au Cercle des Officiers, place Broglie.

Daily program of the Strasbourg 1920 congress showing the postwar atmosphere. (From the proceedings of the 1920 ICM, Eduard Privat 1921.)

On the other hand, there was opposition from certain mathematicians, such as Hardy and Mittag-Leffler, to this exclusion policy. They still were in the minority.

This caused the tone of the congress, as described in the proceedings, to be startlingly postwar: there was a visit to the mausoleum of the Maréchal de Saxe (the Marshal of Saxony); General Taufflied lectured on *La Science en Alsace*; Madame Jeanne Clapier recited the beautiful and inspired ode *Salut à Strasbourg*; there were even special regulations for the lunch of those congressmen who were reserve officers of the Allied Armies. The eagerness to exhibit support for the idea of an Alsatian congress is pervasive.

Even the finances of the congress exhibit the strong nationalistic sentiment surrounding its celebration. The list of donors is three pages long. The largest portion corresponds to the amount donated by companies (ranging from mining companies to brasseries), next is the amount from individual donations, and the smallest part is the direct subvention by the French government.

The peak of this harsh atmosphere came in the closing speeches. Émile Picard, who was president of the IRC, said:

> In respect to certain relations broken by the tragedy of these last years, our successors will determine if a sufficiently long lapse of time and a sincere repentance could allow them to resume some day, and if the ones who excluded themselves from the civilized nations deserve to reenter again. For us, too close to the events, still assume the fine words said by Cardinal Mercier during the war: to pardon certain crimes is to become accomplice with them.

(In judging these words, however, it should be taken into account how the war had destroyed Picard's own family.)

The secretary general of the congress, Koenigs, ended his report to the congress with these words:

> Strasbourg has understood very well the gesture of our mathematician friends in choosing Strasbourg as venue of the first international congress after the war and in organizing it following the feelings of their hearts. They have felt the complex desire of giving

> To Alsace, a testimony of profound affection,
> To others, an example to follow,
> and even to others, a lesson to remember.

Gabriel Koenigs (1858–1931), Secretary General of the UMI. (From the proceedings of the 1920 ICM, Eduard Privat 1921.)

The traditional ceremony of deciding the venue of the next congress at the closing ceremony did not take place this time. Two days before the congress started, delegates from national committees of IRC countries had met in Strasbourg and founded the Union Mathématique Internationale. They had also decided the venue of the next congresses: 1924 in New York and 1928 in Belgium. The day after the meeting, neutral countries were invited to join the union.

It is clear that the hopes of Hardy and Mittag-Leffler were very far from being accomplished in 1920. Mittag-Leffler in particular had very hard feelings, since he insisted that the sixth ICM should take place in Stockholm, as had been decided in 1912 in Cambridge. He did not attend the Strasbourg congress, but he was able, via his Danish colleague Nörlund, to convince the organizers to call it not the International Congress of Mathematicians, but the International Congress of Mathematics, so that it would not interfere with the series of the ICMs. Despite the agreement reached over that point, the proceedings finally bore the name *Congrès International des Mathematiciens*. This issue, however, would cause problems in the future.

Let us forget for a while the dramatic events of those days and enjoy a contribution of the congress to the solving of a deep philosophical problem. The nature of mathematics and the place and final meaning of its truths has been a long standing and controversial issue. From Plato to Kant and the Marxist thinkers, there has always been an intellectual astonishment produced by the "unreasonable effectiveness of mathematics." But in these debates, the working mathematician has participated little. His role has been more that of the miner who, after a tough day deep down in the mine, ends his journey with a diamond in his hand and, coming out to the surface, full of dust and sweat, places the diamond on a white-covered table where several thinkers dressed in white linen debate about gems and their discovery. In his speech at the opening ceremony of the congress, Picard recalled that:

> The geometers like to recall the saying of the great mathematician Lagrange who, one day comparing mathematics with an animal of which everything is

eaten, said "Mathematics is like swine; everything is good."

We should not dispatch too quickly and without careful consideration a commentary by Lagrange, which Picard recalled at an international congress. Rather, let us venture an interpretation of Lagrange's words: In any serious and honest attempt to solve a mathematical problem, there is a faithful look at truth.

Camille Jordan (1838–1922), honorary president of the Strasbourg 1920 congress. (© Collections École Polytechnique.)

TORONTO 1924

THE PHOTOGRAPH of Charles de la Vallée Poussin, president of the Union Mathématique Internationale, presenting a commemorative wreath at the foot of the Soldier's Memorial Tower of the University of Toronto as homage to the students of the university who had laid down their lives in the war, is what best symbolizes the 1924 congress held in Toronto. Then it was announced that a medallion would be inserted into the wall of the tower with the inscription "To the heroes of the University of Toronto, the members of the International Mathematical Congress, Toronto, 1924." This might give the image of a congress fully aligned with the postwar spirit of the Strasbourg 1920 congress, something that is, however, unfair.

If we must believe the official story of the American Mathematical Society, in 1919, the decision to hold the 1920 International Congress at Strasbourg, which the Central Powers were excluded from attending, was taken without consulting the United States and Great Britain (although there are some doubts about the full correctness of this statement). What seems to be unquestionable is that the offer to hold the 1924 congress in New York was made by L. E. Dickson and L. P. Eisenhart, U.S. delegates in Strasbourg, without having consulted their society. By 1922, it was clear that the restrictions on participation imposed by the International Research Council rendered unobtainable any financial backing from the United States. To understand this, we should take into account the strong links that U.S. mathematicians still had with the German academic world, as well as the less dramatic effects of the war on U.S. society.

AN INCIDENT OF COURTESY

Professor Charles de la Vallée Poussin on behalf of the members of the Congress, presenting a commemorative wreath to Sir Robert Falconer, President of the University of Toronto.

The president of the UMI, Charles de la Vallée Poussin, laying a wreath at the foot of the Soldier's Memorial Tower. With him is Sir Robert Falconer, president of the University of Toronto. (From the proceedings of the 1924 ICM, The University of Toronto Press 1928.)

In 1922, the second General Assembly of the International Research Council took place in Brussels. The only possibility for celebrating the 1924 International Congress in the United States was to end the exclusion policy, something that the council was not prepared to accept. As a result, the future of the next congress was in danger. Thus, the offer of John Charles Fields to hold it under the IRC regulations at the University of Toronto was accepted with relief. The details of this episode were concealed and not recorded publicly either at the council's meeting or at the meeting of the International Mathematical Union held during the Toronto congress. In his speech at the opening of the Toronto meeting, de la Vallée Poussin explained that he was happy that the congress was taking place "in the vicinity of New York ... in the New World."

The only indirect mention of this episode can be found in Picard's speech as president of the IRC to the Brussels Assembly in 1922: "Certain Unions have been happy to leave to the International Research Council the trouble of taking some decisions which they are not anxious to take themselves." The Brussels meeting was not a UMI meeting, but the decision to change the venue of the International Congress was taken there. This clearly shows that the independence of the International Mathematical Union from the IRC did not exist.

This long and somewhat cumbersome preamble is needed to understand the level of political interference in this matter, that is, the celebration of the International Congress, which before had depended on purely consensual grounds. So, at the price of not questioning the exclusion policy, Fields had saved the continuity of the ICM series. This has not been Fields' only contribution to the ICM; we will return to this later in "Awards of the ICM."

The Congress met in Toronto by invitation of the University of Toronto and the Royal Canadian Institute, its sessions being held in the buildings of the University. In its organization and the conduct of its proceedings it conformed to the regulations of the International Research Council and the International Mathematical Union.

John Charles Fields (1863–1932), president and main organizer of the 1924 congress. (Courtesy of the Archives of the Mathematisches Forschungsinstitut Oberwolfach.)

Credit page of the 1924 proceedings showing the ruling of the IRC over the congress. (From the proceedings of the 1924 ICM, The University of Toronto Press 1928.)

Fields worked tirelessly to ensure the success of the congress. He traveled across the Atlantic several times; he presided over the organizing committee of the congress, the editorial committee, the finance committee, the committee on transatlantic transportation and European organization, the Canadian national committee of the International Mathematical Union, and the sectional committee on pure mathematics. He was president of the congress and introducer for Section I on Algebra, Number Theory, and Analysis. Indeed, as a consequence of this intense activity, his health suffered: eight years after the congress, he died.

GEOGRAPHICAL DISTRIBUTION

	A	B	C
Argentine	2	–	1
Australia	–	1	–
Belgium	6	2	–
Canada	107	4	7
Cuba	–	1	–
Czechoslovakia	3	–	–
Denmark	3	1	–
Egypt	–	1	–
France	24	18	3
Georgia	1	–	–
Great Britain	58	13	20
Greece	–	1	–
Holland	4	–	2
Hong Kong	1	–	–
India	2	–	–
Irish Free State	2	–	1
Italy	11	3	1
Japan	–	1	–
Jugoslavia	1	–	–
Malta	1	–	1
Mexico	1	–	–
Norway	5	1	–
Peru	–	1	–
Poland	2	1	–
Portugal	2	1	–
Roumania	1	1	–
Russia	4	10	–
Samoa	1	–	–
Spain	3	1	–
Sweden	3	2	–
Switzerland	4	1	–
Ukraine	1	2	–
United States	191	15	64
	444	82	100

A. Delegates and Members present in Toronto.
B. Corresponding Members.
C. Relatives accompanying those listed under A.

Distribution of members of the 1924 congress according to nationality (observe the appearance of a new country: the Irish Free State). (From the proceedings of the 1924 ICM, The University of Toronto Press 1928.)

Despite the unfavorable conditions, Fields was able to organize a successful congress. Participation rose to 444, more than double that of Strasbourg, with 191 participants from the United States, 107 from Canada, and 58 from Great Britain. The number of countries represented was 28, with some historical curiosities: the first appearance of the Irish Free State, with A. W. Conway as delegate of the Irish government and of the National University of Ireland; and the independent presence of three Soviet republics: Georgia, Ukraine, and Russia, with W. Steklov representing the Academy of Sciences of Russia (R.S.F.S.R.). Note that this level of attendance was reached with German, Austrian, Hungarian, and Bulgarian mathematicians still excluded (although two-thirds the of participants were from North America).

With regard to the exclusion policy, Fields was careful not to confront it openly but pointed out the need for its end. Indeed, when preparing the proceedings, he was careful to call the congress the International Mathematical Congress, avoiding the controversial issue of its character as a true international congress of (all) mathematicians, and hence whether or not it was one more in the series of previous ICMs.

The congress was held August 11–16, 1924, at the University of Toronto, with the opening and closing ceremonies in the classical-style Convocation Hall and the lectures in the physics building (see page 159).

As in 1908 in Rome and 1912 in Cambridge, there was a strong presence of the applications of mathematics in the 1924 congress. Fields explained this in the closing of the congress: "The policy of the present Congress was to accentuate more than has been done at previous Congresses the side of applied mathematics."

This can be seen, as usual, in the list of sections where pure mathematics was contained in the first two

University College, University of Toronto. (From the proceedings of the 1924 ICM, The University of Toronto Press 1928.)

sections, followed by a profusion of sections on applications:

* Section I: Algebra, Number Theory, and Analysis,

* Section II: Geometry,

* Section III (a): Mechanics and Physics,

* Section III (b): Astronomy and Geophysics,

* Section IV (a): Electrical, Mechanical, Civil, and Mining Engineering,

* Section IV (b): Aeronautics, Naval Architecture, Ballistics, and Radiotelegraphy,

* Section V: Statistics, Actuarial Science, and Economics,

* Section VI: History, Philosophy, and Didactics.

This applied trend of the congress is also seen in the list of participants, with some unusual people attending. There were participants from companies such as Eastman Kodak, General Electric, American Telephone and Telegraph, Marconi (represented by its President, the Honorable Senator Guglielmo Marconi), from banks and insurance companies, and military personnel from the War Department of the United States and the French Ministries of War and the Navy.

Tuesday, August 14, was devoted to an excursion fully in the applied spirit of the congress. The best way of sharing the joy of such a lovely day is just to read about it in the proceedings of the congress:

The members of the Congress crossed to Niagara, where, on the invitation of the Hydro-Electric Power Commission of Ontario, they inspected the generating station at Queenston. They then proceeded to Niagara

The Toronto power house here stands in the foreground. On the extreme right is seen the intake of the Ontario power house

The members of the 1924 congress visited Niagara Falls and the hydro-electric power station at Queenston-Chippawa. (From the proceedings of the 1924 ICM, The University of Toronto Press 1928.)

Falls, where they were entertained at Luncheon in the Clifton Inn as guests of the Power Commission. After viewing the Falls, and taking the trip along the Gorge Route, the party returned by boat to Toronto.

After the excursion, congress members were entertained in Hart House at a conversazione by the University of Toronto and the Royal Canadian Institute. (There were other congress receptions: the day of the opening, the members of the congress were entertained at a garden party in the York Club by Professor and Mrs. MacLennan, and one afternoon there was a garden party in the Grange organized by the Council of the Art Gallery; afterwards, there was a soirée at the Hunt Club.) This was not the only excursion of this congress; in "Social Life at the ICM," we will discuss the transcontinental excursion that has been (and probably will forever be) the most spectacular of all ICM excursions.

Going back to the Niagara Falls excursion, it is difficult now to imagine the excitement at a power plant, but then countries exhibited with pride these creations of applied science, which assured a future of prosperity and wealth.

Curiously enough, this heavily applied character of the congress was not reflected in the plenary lectures, which were

"La théorie des groupes et les recherches récentes de géométrie différentielle," by Élie Cartan from Paris;

"Outline of the Theory to Date of the Arithmetics of Algebras," by Leonard Eugene Dickson from Chicago;

- *"Considérations sur une équation aux dérivées partielles de la physique mathématique,"* by Jean Le Roux from Rennes;
- "Non-Euclidian Geometry from Non-Projective Standpoint," by James Pierpoint from New Haven;
- *"Sulle operazioni funzionali lineari,"* by Salvatore Pincherle from Bologna;
- *"La géométrie algébrique,"* by Francesco Severi from Rome;
- "Modern Norwegian Researches on the Aurora Borealis," by Carl Størmer from Oslo;
- "Some Characteristics Features of Twentieth-Century Pure Mathematical Research," by William H. Young from London.

The number of communications presented in the sections was very large, 241. Two of them highlight the wide scope of the congress: Maurice Fréchet from Strasbourg lectured on the "Number of Dimensions of an Abstract Set," and Sir Charles A. Parsons, chairman from C. A. Parsons & Co., on "The Steam Turbine."

The historical and bookish side of the congress was covered by the Irish contingent. It was announced that the Royal Irish Academy was contemplating the publication of the collected works of Sir William Rowan Hamilton.

The war was obviously not as present as it had been in the Strasbourg congress—not a surprise since six years had passed since its end. But there were still some traces. One was the laying of the commemorative wreath by de la Vallée Poussin. Another is reflected in the official chosen to represent the government of the Dominion of Canada at the opening session: the Minister of Soldier's Reestablishment (however, surely advised adequately by Fields, he spoke of the event being inaugurated as the International Mathematical Congress, and not of the International Congress of Mathematicians). The strongest reference to the past war was in the speech of de la Vallée Poussin at the opening ceremony, when he explained the meaning of the Strasbourg 1920 congress:

> What then mattered was not only a scientific congress, but a symbol and a feast, the celebration of the liberation of Alsace, and also, as I then said, the liberation of Science from the sacrilegious hands that for so long had used it for their criminal aims.

The opening ceremony shows the control that the International Mathematical Union had over the congresses. Following the established tradition, the clear candidate for president of the congress was Fields, as president of the organizing committee; he was elected. The point was that it was precisely de la Vallée Poussin, as president of the International Mathematical Union, who made the proposal and not the president of the Strasbourg congress or his representative.

Memorial Tower, University of Toronto

The University of Toronto. (From the proceedings of the 1924 ICM, The University of Toronto Press 1928.)

First group photograph of an ICM: Toronto 1924. (From the proceedings of the 1924 ICM, The University of Toronto Press 1928.)

On Friday, August 15, the second meeting of the International Mathematical Union since its founding took place. There, Salvatore Pincherle was elected its new president. The rest of the meeting is best explained by G. H. Hardy (who did not attend the congress as a protest against the exclusion policy). Referring to the "boycott of ex-enemy nations organized by the 'International Research Council,'" in a note published after the congress he explained that:

> Many American mathematicians who attended the Congress discovered for the first time when they arrived at it that Germans were excluded. A good deal of indignation was expressed, and the representatives of the American Mathematical Society moved a resolution for the removal of the ban.

The proposal was supported by Denmark, Great Britain, Holland, Italy, Norway, and Sweden. Its text (not included in the minutes of the meeting) was:

> The American Section of the International Union requests the International Research Council to consider whether the time is ripe for the removal of restrictions on membership now imposed by the rule of the Council.

This situation resulted in the withdrawal of the official proposal to hold, under the IRC regulations, the next congress in Brussels. For the first time in the history of the ICM, the decision on the venue of the next congress was not taken (a similar event occurred, but for completely different reasons, in the 1958 congress in Edinburgh).

The decision was delayed until 1926 to be considered by the Governing Committee of the union. The protest of the international mathematical community against the exclusion policy was gaining strength.

We have a precious keepsake of this congress: the first group photograph of all participants in an ICM (there are only five of this type of group photograph). It was taken on Monday, August 11, after the opening ceremony in front of the Physics building. It shows an unusually old-fashioned group of people.

BOLOGNA 1928

It makes me very happy that after a long, hard time all the mathematicians of the world are represented here. That is as it should be and as it must be for the prosperity of our beloved science.

Let us consider that we as mathematicians stand on the highest pinnacle of the cultivation of the exact sciences. We have no other choice than to assume this highest place, because all limits, especially national ones, are contrary to the nature of mathematics. It is a complete misunderstanding of our science to construct differences according to peoples and races, and the reasons for which this has been done are very shabby ones.

Mathematics knows no races ... For mathematics, the whole cultural world is a single country.

Today these words of David Hilbert still symbolize the universality of mathematics, but when they were pronounced, they also represented the end of a dark era in the international collaboration in mathematics. The 1928 International Congress held in Bologna marked the end of the exclusion policy established after the Great War and the return to the original spirit of the international congresses as conceived in Zurich in 1897.

However, the hero of the congress was Salvatore Pincherle from Bologna. Paradoxically, he was, at the time, president of the International Mathematical Union but had to fight against the union and the Research Council to have the 1928 congress open to mathematicians from all nations.

Salvatore Pincherle (1853–1936), president of the 1928 Congress and president of the UMI. (Courtesy of the Unione Matematica Italiana.)

The proceedings of the 1928 congress devote six pages to explaining in detail this important episode in the history of international scientific cooperation. The sequence of events was the following. After the signing in 1926 of the Treaty of Locarno (where the renunciation of resorting to war was agreed upon), Germany entered the League of Nations. Then, the Research Council, with Picard as president, changed its statutes allowing the former Central Powers to join the Council and

The 1928 congress took place under the auspices of the University of Bologna in order to allow the participation of mathematicians from the Central Powers (the image is of the Istituto Chimico, where the lectures were held). (Courtesy of the Archivio Storico, Università di Bologna.)

then inviting them to do so. However, the existence in Germany of not one but several scientific academies (those of Berlin, Göttingen, Leipzig, and Munich) caused an uncertain situation, and there was no immediate response to the invitation. It was at this stage that Bologna was chosen by the International Mathematical Union as the venue of the 1928 congress. The organizing committee, with Pincherle as president, reinstated the old tradition of sending an open invitation to all mathematicians and mathematical societies. The announcement reached the Deutsche Mathematiker-Vereinigung, who inserted it in its *Jahresberichte*.

Opposition against attending the Bologna congress arose inside Germany, led mainly by Ludwig Bieber-

bach from Berlin. This was effectively counteracted by a strong statement of Hilbert in favor of participating in the congress. On the other hand, the secretary of the International Mathematical Union, Koenigs, in agreement with the president of the Research Council, Picard, communicated that under the conditions of free attendance, the congress would be considered illegally convened. At the same time, some important mathematicians, along with many countries and societies, had said that unless the congress was open to all mathematicians, they would abstain from attending. Pincherle's formal solution to this impossible puzzle was to put the congress under the patronage of the University of Bologna, thus avoiding the International Mathematical Union's control and maintaining the open invitation for attending.

What then happened was a true test of the opinion of the mathematical community. The congress opened on September 3, 1928, in Bologna, with 836 participants coming from 36 countries. These were the largest figures ever seen before for an ICM. (For this congress, the proceedings comprised seven volumes!) The largest national group was, obviously, Italians with 336 participants, but the second and third were Germans, with 76, and French, with 56. There had been no effective boycott.

	Effettivi	Persone di Famiglia	Totale
Argentina	7	—	7
Austria	9	1	10
Belgio	10	4	14
Brasile	2	—	2
Bulgaria	5	—	5
Canadà	4	3	7
Cecoslovacchia	15	7	22
Danimarca	9	5	14
Egitto	2	—	2
Finlandia	2	—	2
Francia	56	35	91
Germania	76	30	106
Giappone	11	2	13
Grecia	8	2	10
Guatemala	1	2	3
India	5	2	7
Inghilterra	47	17	64
Irlanda	2	—	2
Italia	336	76	412
Jugoslavia	4	1	5
Lettonia	1	—	1
Lituania	1	1	2
Norvegia	8	4	12
Olanda	9	6	15
Palestina	6	—	6
Polonia	31	10	41
Romania	11	8	19
Russia	27	3	30
Serbia	1	—	1
Spagna	11	1	12
Stati Uniti d'America	52	24	76
Svezia	4	3	7
Svizzera	29	19	48
Turchia	2	2	4
Ucraina	10	3	13
Ungheria	22	9	31
	836	280	1116

Distribution of members of the 1928 congress according to nationality. (From the proceedings of the 1928 ICM, Zanichelli 1928.)

The entrance of Hilbert leading the German delegation into the hall of the old *Archiginnasio di Bologna* (see page 160) for the opening ceremony of the congress is legendary. Germans had been absent from the international congresses since the war. Constance Reid, in her book on Hilbert, recreates those emotive moments: "For a few minutes there was not a sound in the hall. Then, spontaneously, every person present rose and applauded."

A great deal of the applause was directed at Hilbert as a living incarnation of the spirit of cooperation of the ICM. But part of the applause can be interpreted as reflexive: the international mathematical community had won. The will for collaboration had overcome all difficulties, and the community was happy and proud to be united again.

Hilbert's entrance into the 1928 congress leading the German delegation was met by a standing ovation. (Courtesy of the Niedersächsische Staats- und Universitätsbibliothek Göttingen. Sammlung Voit: D. Hilbert, Nr. 17.)

Very symbolic of the renewed spirit of the congress was placing Hilbert's lecture *"Probleme der Grundlegung der Mathematik"* as the first plenary lecture.

Lecture room of the Istituto Chimico where Hilbert lectured in the Bologna 1928 congress. (Courtesy of the Archivio Storico, Università di Bologna.)

Corresponding to the increase in participation, the number of plenary lectures also increased: to sixteen! This caused a very intense schedule of plenary sessions: daily, but for the day of the congress excursion, there were three consecutive plenary lectures. They were:

* *"Le equazioni differenziali della dinamica economica,"* by Luigi Amoroso from Rome;
* *"Le calcul des probabilités et les sciences exactes,"* by Émile Borel from Paris;
* *"La geometria algebraica e la scuola italiana,"* by Guido Castelnuovo from Rome;
* *"L'analyse générale et les espaces abstraits,"* by Maurice Fréchet from Strasbourg;
* *"Le développement et le rôle scentifique du calcul fonctionnel,"* by Jacques Hadamard from Paris;
* *"Mathematische Probleme der modernen Aerodynamik,"* by Theodore von Kármán from Aachen;
* *"Sur les voies de le théorie des ensembles,"* by Nikolai N. Lusin from Moscow;
* *"Leonardo da Vinci nella storia della matematica e della meccanica,"* by Roberto Marcolongo from Naples;
* *"Le bonifiche in Italia,"* by Umberto Puppini from Bologna;
* *"Il contributo italiano alla teoria delle funzioni di variabili reali,"* by Leonida Tonelli from Bologna;
* "Differential Invariants and Geometry," by Oswald Veblen from Princeton;
* *"La teoria dei funzionali applicata ai fenomeni ereditari,"* by Vito Volterra from Rome;
* *"Kontinuierliche Gruppen und ihre Darstellungen durch lineare Transformationen,"* by Hermann Weyl from Zurich;
* "The Mathematical Method and Its Limitations," by William H. Young.

(If the reader has counted the lectures, he/she should have found that there is one missing; the reasons will be provided later.)

A similar growth factor affected the sections and the communications presented. The proceedings record that 419 communications had been received, but only 330 appeared in print. The list of sections had the traditional beginning with some permutation of the fields of pure mathematics, and the classical ending with historical, philosophical, and educational issues. Where there was an increase was in the sections devoted to applications of mathematics:

- Section III: Mechanics, Astronomy, Geodesics, Mathematical Physics, and Theoretical Physics,
- Section IV: Statistics, Mathematical Economy, Calculus of Probability, and Actuarial Science,
- Section V: Engineering and Industrial Applications.

The explicit appearance of the calculus of probability was a clear sign of the strength of an emerging field. Indeed, in Section IV, a resolution was approved expressing "the wish that the papers concerning the Theory of Probability and its Applications should be as far as possible concentrated in a few journals."

The members of the 1928 congress were presented a book on the mathematical school of Bologna. (Courtesy of Zanichelli editore.)

The members of the 1928 congress were presented a book on newly discovered chapters of Bombelli's algebra. (Courtesy of Zanichelli editore.)

The congress also reinstated another of its old traditions and was rich in books given as gifts to participants: a scientific memoir expressly prepared by Luigi Bianchi; the last issue of the journal *Annali di Matematica Pura ed Applicata*, founded in 1850; a treatise on the mathematical school of Bologna by Ettore Bortolotti; and the *Preface to the Unedited Books of the Algebra of Rafael Bombelli*. All this was done in collaboration with the editorial house Nicola Zanichelli, who was deeply involved in the congress and who had produced the printed matter for the congress.

The historical side of the congress was devoted to great figures of Italian mathematics. On the afternoon of Sunday, September 9, the congress participants went to the family house of Scipione del Ferro, where a commemorative stone slab was placed. Afterwards they proceeded to the church of the Mascarella, where Bonaventura Cavalieri had been prior, to place another commemorative stone.

The language issue was very liberally treated in this congress; five languages were used for the invitations and lectures: Italian, French, German, English, and Spanish—and also Latin, classical or *sine flexione*. Indeed, the Rector of the University of Bologna read his three-page speech to the congress in Latin.

JOSEPHI ALBINI, Rector Athenaei bononiensis. Verba in Archigymnasio habita III Non. Sept. A. MCMXXVIII quo die primum sedit Mathematicorum ab omni gente conventus.

Vobis omnibus, Viri clarissimi, qui convenistis pervetus haec Disciplinarum et Artium Universitas non modo laeta lubens obviam occurrit sed magnas etiam agit gratias, quod invitanti sibi tot et tales estis obsecuti. Per vos enim fit ut undique fere coisse videantur eximii qui Mathematicen excolunt vel ipsam vel in aptis haustisve ex ipsa artibus et facultatibus, atque hic unanimi, iis intermissis quibus generatim moveri possint studiis, id unum spectent quob habent universi doctrinae et veritatis propositum. Tanta igitur inest et consiliis vestris et huic concilio auctoritatis atque humanitatis laus, ut iure meritoque favere et interesse censuerint et Civitas magistratu suo cum signis avitae libertatis ac fascibus receptis et Is qui fortiter feliciter praeest rebus patriis summum legans Disciplinae publicae Ministrum et ipsa Sabaudica Domus, regium nomen, omen Italicum, adstante hac lectissimi Principis iuventa pro Rege optimo eodemque doctissimo.

The Rector of the University of Bologna delivered his speech to the 1928 congress in Latin. (From the proceedings of the 1928 ICM, Zanichelli 1928.)

A closer look at the list of participants shows interesting facts. There were 72 women participating in the congress; of them, 23 were not Italian. Among these was Emmy Noether, who delivered the lecture *"Hypercomplexe Grössen und Darstellungstheorie in arithmetischer Auffassung."* We find participants from Latvia, Lithuania, the University of Jerusalem, the Polytechnical Institute of Haifa in Palestine, and the Deutsche Universität in Prague.

There was some attention paid to the International Commission on Mathematical Instruction, which had been dissolved in 1920 by the International Mathematical Union. Since 1909, it had produced 291 reports resulting in more than 13,000 written pages. The congress decided to reconstitute the commission, appointing David Eugene Smith from New York as president (the former president, Felix Klein, had died in 1925).

We have already seen that the 1904 congress in Heidelberg was the great congress of the German empire. In a similar way, the Bologna congress was a great international exhibition for the Fascist state of Benito Mussolini. All throughout the report of the congress, the strong presence of the political regime is evident. Naturally, Mussolini was president of the honor committee for the congress. Another member of that committee was the secretary general of the National Fascist Party; the congress enjoyed the hospitality of the *Casa del Fascio*. The congress was greeted in the name of the "Fascist Government," and an important part of the financing of the congress came from several of the Fascist labor unions.

COMITATO D' ONORE

Presidente: S. E. Cav. BENITO MUSSOLINI, Capo del Governo.
S. E. T. TITTONI, Presidente del Senato.
S. E. A. CASERTANO, Presidente della Camera dei Deputati.
S. E. A. TURATI, Segretario del Partito Nazionale Fascista.
S. E. P. BADOGLIO, Maresciallo d'Italia, Capo di Stato Maggiore Generale.

Sign of the times: The Honor Committee of the Bologna 1928 congress was presided over by Benito Mussolini. (From the proceedings of the 1928 ICM, Zanichelli 1928.)

There was a tremendous effort to please the congress participants, which was laudable but perhaps a bit overdone. The facilities granted by the state were overwhelming: discounts, sometimes up to 50 percent, on train and ship tickets, hotels, and restaurants; free travel on trams; and free entrance to museums and art galleries (obtained by showing the congress badge, an artistic medal carried out by the engraver Johnson following a design of Professor Borghesani).

Even the presence of the Archbishop of Bologna at the opening ceremony was an indication of political influence (that had not occurred in the 1908 congress in Rome); Mussolini was close to signing the first official agreement between the Catholic Church and the Italian state. The selection of the official languages of the congress can also be seen from this viewpoint. The inclusion of Spanish as an official language could only be interpreted as a political gesture in support of the political regime in Spain at the time (an authoritarian military dictatorship).

The speech of the mayor of Bologna was a clear example of all that atmosphere: "The Fascist Bologna is proud to offer its hospitality and to exhibit what Bologna has become under the vivifying impulse of Fascism."

The closing session was magnificent. It took place not in Bologna but in Florence, in the Cinquecento Hall of the renowned Palazzo Vecchio. It was reported that "there was a pageant with trumpeters, soldiers, and heralds in ancient costumes." The mayor of Florence delivered a speech, unusually beautifully written. He praised the glories that Tuscany had given to mathematics: in the thirteenth century, Leonardo Fibonacci wrote his book *Liber abbaci*; in the fourteenth century, Raffaello Canacci published the first treatise on algebra written in a modern language (that is, not in Latin); in the fifteenth century, Leonardo da Vinci devised the new mechanics; and Galileo Galilei in the sixteenth century founded modern astronomy and rational mechanics. But the contributions were not only to mathematics. As in classical Greece, art combined with science, and Tuscany had given the world artists such as Botticelli and Raphael. He spoke of the libraries of Tuscany, rich in treatises on arithmetic, algebra, and geometry, all written in the fourteenth and fifteenth centuries by the merchants who were bankers for Popes, emperors, and kings of Europe.

This was the proper speech for what occurred next, the last plenary lecture delivered by George David Birkhoff, from Cambridge, Massachusetts, on *"Quelques éléments mathématiques de l'art."*

After the closing session, there was a reception at the City Hall. Congress participants were later invited to visit an exhibition of rare books on the mathematical sciences in the Palazzo Vecchio, organized by the directors of the Biblioteca Nazionale and the Biblioteca Laurenziana. This was something not to be missed!

After the Bologna congress, what future could be expected for the international congresses and the International Mathematical Union? The day before the closing of the congress, there was a meeting of the union. Its character was necessarily unofficial since the secretary general had opposed convoking it. The president of the union, Salvatore Pincherle, was congratulated for his management of the situation, for the success of the congress; he was supported in all of his decisions. Nevertheless, he decided to resign, four years before the end of his mandate. There was general agreement on the necessity of reconsidering the situation of the union. For the next congress, there were proposals from Holland and Prague, but neither had the backing of the majority. Thus, it was decided to propose a neutral solution, that is, Switzerland. The next day, at the end of the closing ceremony, the location of the next ICM was unanimously approved by the congress participants. The congress would take place in Zurich in 1932.

ATTI
DEL
CONGRESSO INTERNAZIONALE
DEI MATEMATICI
BOLOGNA 3-10 SETTEMBRE 1928 (VI)

The controversy over the validity of the 1920 and 1924 ICMs appears again: the Bologna 1928 congress is labeled the sixth ICM, following the fifth in Cambridge in 1912. (From the proceedings of the 1928 ICM, Zanichelli 1928.)

The International Mathematical Union had lost its grip on the congresses. This was patently clear, as evidenced in the name of the congress imprinted in the proceedings: the name was again International Congress of Mathematicians. But the key issue was the number: VI; that is, the next number after the 1912 ICM in Cambridge. The congresses of Strasbourg in 1920 and Toronto in 1924 were not considered true ICMs and were removed from the list of the congresses.

ZURICH 1932

Tʜᴇ ɪᴍᴀɢᴇ ᴏғ ᴛʜᴇ German mathematician Ludwig Bieberbach from Berlin and the Polish mathematician Wacław Sierpiński from Warsaw on the steps of the Eidgenössiche Technische Hochschule, ETH, of Zurich symbolizes perfectly the 1932 international congress. Bieberbach, who had opposed German participation in the 1928 Bologna congress, was invited to give a plenary lecture in Zurich. The 1932 ICM was a smooth congress for the mathematical community, which found itself reunited and in harmony.

The tensions caused by the Great War and the political interference in the international congresses had passed. The political situation in Europe was, for the moment, stable. The Nazis had not yet risen to power in Germany. (In later years, Bieberbach became a Nazi supporter and the main promoter of the journal *Deutsche Mathematik*.) Nothing foreshadowed the dark clouds that within a short time would descend over Europe and the world.

The only threat was the deep economic crisis of the depression of 1929. Its presence was felt everywhere throughout the congress: Rudolf Fueter from Zurich, in his opening speech as president of the organizing committee, explained that they had been able to organize a simple but dignified congress despite the "difficult times;" Oswald Veblen from Princeton, in his speech as president of the U.S. delegation, praised the effort to continue with the invitation to host the

Bieberbach   Sierpinski

The German mathematician Ludwig Bieberbach and the Polish mathematician Wacław Sierpiński on the steps of Zurich's ETH during ICM 1932. (Courtesy of the Bildarchiv der ETH-Bibliothek Zürich.)

congress despite the difficult economic situation; Hermann Weyl from Göttingen referred to the "terrible economic depression" and the number of international congresses that had recently been canceled.

The effects of the economic crisis are also seen in the finances of the congress. More than half of the donations came not from the Swiss Confederation, the

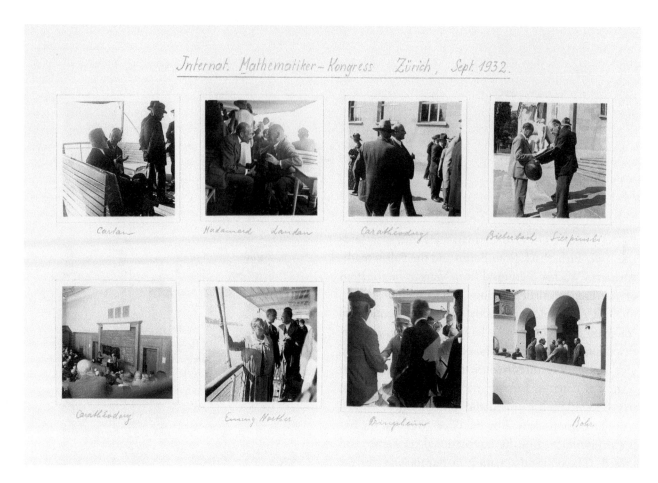

Internat. Mathematiker-Kongress Zürich, Sept. 1932.

Cartan — Hadamard, Landau — Carathéodory — Bieberbach, Sierpinski

Carathéodory — Emmy Noether — Dringsheim — Bohr

Board with daily photos from the 1932 congress. (Courtesy of the Bildarchiv der ETH-Bibliothek Zürich.)

Zurich Canton, or the City of Zurich, but from private companies, mainly banks and insurance companies. A special donation was offered by "several music lovers" for the concert at the Tonhalle in homage to the congress. Curiously enough, Fueter observed that the fees were very similar to those of the 1897 congress: 25 francs in 1897 and 30 in 1932 (and for both congresses, 15 francs for accompanying family members).

Besides the invitation to Bieberbach, the Swiss organizers sent signals of reconciliation in all directions. Following the old traditions reestablished in the Bologna congress, the invitation to attend the congress was sent to all academies, mathematical societies, and mathematicians in the world. At the opening session, held on Monday, September 5, in the Auditorium Maximum of the ETH, a decision was made to send a telegram to Émile Picard as a sign of admiration and respect; Picard answered some days later. Another sign of respect was given to Hilbert, who presided over the first plenary session. The congress honored him by standing up at his entrance into the hall. Tradition was reflected in the presence of 90-year-old Karl F. Geiser, president of the first international congress.

The most telling sign of the reconciliation, with which the whole mathematical community was involved, came with the address of Hermann Weyl, president of both the Deutsche Mathematiker-Vereinigung and the German delegation, delivered at the reception offered by the Swiss Government. He said:

> Here we attend to an extraordinarily improbable event. Given the number n corresponding to the recently opened International Congress of Mathematicians, we have the inequality $7 \leq n \leq 9$; unfortunately our axiomatic foundations are not sufficient to give a more precise statement.

This was an elegant and polite solution to the controversial issue of the numbering of the congresses. This congress would be the ninth if the Strasbourg and Toronto congresses were considered as true international congresses; otherwise, it would be the seventh. Weyl's proposal was implicitly accepted: ever since, the ICMs have not been numbered!

The 1932 congress showed several features of the modern congresses. One of these was in its scientific content. The number of plenary lectures increased significantly. This was necessary; the research community was larger, and the ICM had gained acceptance and popularity. The proceedings of the congress presented 20 lectures, which, in chronological order of their delivery, were the following:

- "Idealtheorie und Funktionentheorie," by Rudolf Fueter from Zurich;
- "Über die analytischen Abbildungen durch Funktionen mehrerer Veränderlicher," by Constantin Carathéodory from Munich;
- "Essai sur le développement de la théorie des fonctions de variables complexes," by Gaston Julia from Paris;
- "Die Aufgaben der modernen Galoisschen Theorie," by N. Chebotaryov from Kazan;
- "Sur la théorie des équations intégrales linéaires et ses applications," by Torsten Carleman from Stockholm;
- "Les espaces riemanniens symétriques," by Élie Cartan from Paris;
- "Operationsbereiche von Funktionen," by Ludwig Bieberbach from Berlin;
- "The Calculus of Variation in the Large," by Marston Morse from Cambridge, MA;
- "Hyperkomplexe Systeme in ihrem Beziehungen zum kommutativen Algebra und zur Zahlentheorie," by Emmy Noether from Göttingen;
- "Fastperiodische Funktionen einer komplexen Veränderlichen," by Harald Bohr from Copenhagen;
- "La théorie générale des fonctions analytiques de plusieurs variables et la géométrie algébrique," by Francesco Severi from Rome;
- "Über die Riemannsche Fläche einer analytischen Funktion," by Rolf Nevanlinna from Helsingfors;
- "L'aspect analytique du problème des figures planétaires," by Rolin Wavre from Geneva;
- "Some Problems in Topology," by James W. Alexander from Princeton;
- "Sur l'existence de la dérivée des fonctions d'une variable réelle et des fonctions d'intervalle," by Frédéric Riesz from Szeged;
- "Le théorème de Borel-Julia dans la théorie des fonctions méromorphes," by Georges Valiron from Paris;
- "Sur les ensembles de points qu'on sait définir effectivement," by Wacław Sierpiński from Warsaw;
- "Sur les liaisons entre les grandeurs aléatoires," by Sergi Bernstein from Kharkov;
- "Neuere Methoden und Probleme der Geometrie," by Karl Menger from Vienna;
- "Anschauung und Denken in der klassischen Theorie der griechischen Mathematik," by J. Stenzel from Kiel.

There was actually one more plenary lecture delivered but, at the special request from the lecturer, not included in the proceedings: *"Mathematische Methoden der Quantenmechanik,"* by Wolfgang Pauli from Zurich. The total number of planned plenary lectures was 22 because there was one more lecture scheduled, which, for some unexplained reason, was not delivered. It was Hardy's "Recent Work in Additive Theory of Numbers" (Hardy did attend the congress and presented a communication with Littlewood). Of all the plenary lectures, five were one-hour addresses (the ones by Fueter, Carathéodory, Cartan, Severi, and Stenzel), and the rest lasted half an hour.

Børge Jessen, from Copenhagen, lecturing at the 1932 congress. (From a 1932 Swiss newspaper.)

While plenary sessions took place at the ETH, the communications presented to the sections were delivered in the lecture rooms of the University of Zurich. The main reason for this was the large number of sections:

* Section I: Algebra and Number Theory,

* Section II: Analysis,

* Section III: Geometry,

* Section IV: Calculus of Probability, Actuarial Mathematics, and Statistics,

* Section V: Technical Mathematical Sciences and Astronomy,

* Section VI: Mechanics and Mathematical Physics,

* Section VII: Philosophy and History,

* Section VIII: Pedagogy.

Additionally, in Section II there were three parallel sessions; in Sections III and IV there were two parallel sessions. There were 12 groups meeting simultaneously! A total of 247 communications were printed in the proceedings, with the Analysis section being the one with the most papers—81. Section VII hosted a meeting of the International Commission on Mathematical Instruction, and, as a result of the work of the section, the congress invited the commission to continue its work (although it remarked that it assumed no financial obligations). Jacques Hadamard was appointed as president of the commission until the 1936 congress.

Another modern feature of the congress was the typography used to print the proceedings, neat, straight, and clean. Compare it with the Gothic style used in some Swiss newspapers at the time.

INTERNATIONALER
MATHEMATIKERKONGRESS
ZÜRICH 1932

Der internationale Mathematikerkongreß in Zürich.

Typesetting of the 1932 congress proceedings (top) and of a local newspaper (bottom). (From the proceedings of the 1932 ICM, Orell Füssli 1932.)

The traditional exhibition of current mathematical books also showed Swiss mathematical instruments; it was organized by J. J. Burckhardt, of whom we have spoken in the preface of this book. The exhibition took place at the ETH. Each participant received a

Exhibit of calculating machines and instruments at the 1932 congress. (Courtesy of J. J. Burckhardt.)

commemorative volume of the Swiss journal *Commentarii Mathematici Helvetici* as a gift.

The social side of the congress included a gathering organized by the Swiss state for mathematicians and guests in the Stadttheater. There was also a festival with a buffet and dancing, and a tea at the Grand Hotel Dolder offered by the City of Zurich.

A close look at the list of participants and delegates to the congress always reveals interesting features (see page 97). For this congress, the Russian delegation was composed of ten mathematicians, a considerable decrease from the 27 who had participated in

the Bologna 1928 congress. It is interesting to see the opinions at the time regarding the trend that might be taken by Soviet mathematics. D. E. Smith, reporting on the congress for the journal *Science,* commented: "As to Russia there has been a rather popular feeling in other countries that she is concerned only with the immediate applications of mathematics to the industrial field." But he concluded that a look at the papers presented by the Russian contributors "is sufficient to show that the subject is not looked upon by the Soviet states as merely utilitarian in the narrow sense." In the list of delegates to the congress, we also find some

Daily life at the 1932 congress. (Courtesy of J. J. Burckhardt.)

Group photograph of the 1932 congress. (Courtesy of Antal Varga, Szeged.)

"White Russian" institutions: Groupe Académique Russe de Paris, Institut Supérieur Technique Russe en France, and the Groupe Académique Russe en Yougoslavie. We also find delegates from the Hebrew University of Palestine.

As was the case for the Toronto congress, a group photograph of the whole congress (853 participants including the accompanying members) was taken on Wednesday, September 7, at 2:30 p.m. at the entrance of the building of the University of Zurich, where the communications had been delivered. Only eight years had passed since the Toronto congress, but the change can be appreciated in the picture: faces, attitudes, fash-

ion, the number of women in the picture. The photographs seem to involve people of different centuries.

The presence of women in this congress was important, not because of the number, only 34, but for their role. Emmy Noether gave a plenary lecture. This was the first time that a woman had been invited to do so at an international congress (see page 97). The next woman had to wait 58 years until the Kyoto 1990 congress. There were also five women who presented communications in the sections: Mary L. Cartwright from Cambridge and Odette Deisme from Le Havre, in Section II; and in Section III, Louise Cummings from New York, Ingebrigt Johansson from Oslo, and Marie

Charpentier from Poitiers. The International Federation of University Women and the American Association of University Women sent official delegates to the congress.

Emmy Noether

Emmy Noether (1882–1935), the first woman to be a plenary lecturer at an ICM (the next was in 1990 in Kyoto). (Courtesy of the Bildarchiv der ETH-Bibliothek Zürich.)

One of the highlights of the congress was the proposal of John Charles Fields to create an award for young mathematicians. Fields had died a month before the congress began, so it was John L. Synge, secretary of the Toronto 1924 congress, who, at the opening session, introduced the award as a Canadian contribution to the cause of international scientific cooperation and provided precise details of the award. The congress committee agreed to consider Fields' proposal and announce its decision at the closing session of the congress. The issue of awards had always been a controversial one, but in this case it represented a step forward in the internationality of the congresses. At the closing session, at the motion of the president, the congress accepted the following proposal:

The International Congress of mathematicians held in Zurich accepts with thanks the offer made by the late professor Fields of two medals to be awarded to two mathematicians at intervals of four years by the International Congresses.

The executive committee, in accordance with the memorandum of professor Fields, chooses a small committee consisting of the following gentlemen: Birkhoff, Carathédory, Cartan, Severi, Takagi.

Verteilung der Kongressmitglieder auf die Staaten

	Teilnehmer	Begleiter		Teilnehmer	Begleiter
Ägypten	5	2	Übertrag	353	87
Afrika	2	1	Lettland	1	1
Belgien	7	1	Mexiko	1	—
Bulgarien	3	1	Norwegen	9	5
Canada	2	—	Österreich	10	—
China	3	3	Palästina	2	—
Dänemark	6	2	Persien	1	—
Deutschland	118	24	Polen	20	2
England	37	12	Portugal	2	1
Finnland	3	—	Rumänien	7	5
Frankreich	69	20	Russland	10	2
Griechenland	5	—	Schweden	5	2
Holland	16	1	Schweiz	144	41
Japan	3	1	Spanien	10	—
Indien	2	—	Tschechoslowakei	12	1
Irland	4	2	Türkei	2	—
Italien	64	17	Ungarn	12	3
Jugoslavien	4	—	U. S. A.	66	36
Übertrag	353	87	Total	667	186

Distribution of members of the 1932 congress according to nationality. (From the proceedings of the 1932 ICM, Orell Füssli 1932.)

The attitude of the International Mathematical Union regarding the Bologna congress condemned its future existence. After some debate, a new president was appointed. He was William Henry Young, the retired president of the London Mathematical Society. The International Research Council no longer existed. It had been transformed into another nonaggressive organization. All but one of the scientific unions, the Union Mathématique Internationale, had adapted their statutes to the new situation. The statutes of the union had expired in 1931, and no action was then taken to renew them. There was a meeting of the union in Zurich, coinciding with the congress (not even mentioned in the proceedings of the congress). Many national delegations were extremely skeptical about the existence of the union (the U.S. delegation considered that "a permanent international organization had no

INTERNATIONALER
MATHEMATIKERKONGRESS
ZÜRICH 1932

I. TEILNEHMERLISTE

MIT DEN ZÜRCHER ADRESSEN · AVEC LES ADRESSES DE ZURICH

BUCHDRUCKEREI NEUE ZÜRCHER ZEITUNG, ZURICH

INTERNATIONALER
MATHEMATIKERKONGRESS
ZÜRICH 1932

II. TEILNEHMERLISTE

MIT DEN ZÜRCHER ADRESSEN · AVEC LES ADRESSES DE ZURICH

BUCHDRUCKEREI NEUE ZÜRCHER ZEITUNG, ZURICH

Booklet listing members of ICM 1932. (Courtesy of J. J. Burckhardt.)

problems important enough to warrant its existence"). It seems that a nonunanimous decision was taken to liquidate the union. In this regard, the congress resolved that:

> An international commission is formed in order to re-study the question of the international collaboration in the sphere of mathematics and to make propositions with regard to its organization at the next congress.

The commission was presided over by Francesco Severi (Italy), and its members were P. Alexandrov (U.S.S.R.), H. Bohr (Denmark), L. Fejér (Hungary), G. Julia (France), J. L. Mordell (Great Britain), E. Terradas (Spain), Ch. de la Vallée Poussin (Belgium), O. Veblen (U.S.A.), H. Weyl (Germany), and S. Zaremba (Poland).

A real sign of the dissolution of the union was that it was the congress that accepted the invitation to hold the next congress in Oslo, presented by Alf Guldberg in the name of the Norwegian mathematicians. And, as had already occurred in 1912, there was a proposal from Athens for the following congress.

Michel Plancherel, Rector of the ETH, at one of the official receptions of the congress, expressed some hopes that to many may have sounded naive at the time:

> Between the two Zurich congresses war has passed over spreading hatred, accumulating ruins and using science for its destructive aim. Perhaps in some decades Zurich will host for a third time the most selected of geometers of the whole world. I hope that whoever is in this post of mine will not have to evoke again such a specter.

And the worries were indeed naive, but not in relation to Zurich hosting a third international congress, which, in fact, it did in 1994, but because they had no idea that the specter of war was soon to be evoked.

We end with a poem that Hermann Weyl read at one of the official receptions. The poem was written by the eighteenth-century Swiss savant Albrecht von Haller, who worked at the University of Göttingen. The poem was dedicated to the canon and mathematician Gessner from Zurich:

Bald steigest du auf Newtons Pfad
In der Natur geheimen Rath,
Wohin dich deine Meß-Kunst leitet.
O Meß-Kunst, Zaum der Phantasie!
Wer dir will folgen, irret nie;
Wer ohne dich will gehn, der gleitet.

Translating poetry is a truly impossible task, but in any case it should be tried (I hope German-speaking readers will forgive me for this):

Soon you ascend through Newton's path
To Nature's private council,
Where you arrived by measuring-art.
Oh measure, reigning in imagination!
Who follows you shall never fail;
Who shuns you will be lost in dark.

OSLO 1936

At the closing session, the 12 of September of 1932, the International Congress of Mathematicians of Zurich decided to accept the legacy of the late Professor Fields which allowed the awarding, at each international congress, of two gold medals to two young mathematicians acknowledged for particularly remarkable works. At the same time, a commission was named to designate the two laureates of the Oslo congress. It was composed of Mr. Birkhoff, Mr. Carathéodory, Mr. Cartan, Mr. Severi, and Mr. Takagi. This commission was presided over by Mr. Severi who, not having been able to attend the Oslo congress, has asked me to replace him in the presidency. The commission has come to this agreement of designating Mr. Lars Ahlfors from the University of Helsinki and Mr. Jesse Douglas from the Massachusetts Institute of Technology as the two first awardees of the Fields Medals. Mr. Carathéodory has agreed to report on the work of the two laureates; he will read his report.

T HESE WORDS OF ÉLIE CARTAN, acting president of the Fields Committee, at the opening ceremony of the Oslo 1936 congress inaugurated a new feature of the congress protocol, the awarding of the Fields Medals. The congress was opened on Tuesday, July 14, in the Aula of the University of Oslo in the presence of King Haakon VII of Norway. This magnificent hall was adorned with huge murals by the Norwegian painter Edvard Munch. There is a well-known photograph of the ceremony, with the King in the aisle and several well-known mathematicians in the front rows (second and fourth in the first row are, respectively, Cartan and Carathéodory; in the second row Ahlfors and Wiener are, respectively, first and second).

After the report by Carathéodory on the work of the laureates, Cartan presented the medals. Douglas' medal was collected by Norbert Wiener in Douglas' name because, as it is explained in the proceedings of the congress, "Mr. Douglas was absent from the ceremony and was not able to receive himself the medal to him assigned." (Douglas was in Oslo; the precise cause of his nonattendance is still one of the mysteries of the

The press reported the awarding in the 1936 congress of the first Fields Medals to Lars V. Ahlfors and Jesse Douglas (Norbert Wiener accepted the medal in Douglas' name). (Courtesy of Nils Voje Johansen.)

Opening ceremony of the 1936 congress at the Aula of the University of Oslo. King Haakon VII of Norway is in the aisle; second and fourth in the front row are Cartan and Carathéodory, respectively, members of the Fields Committee; Ahlfors and Wiener are in the second row. (Courtesy of the National Library of Oslo NBO Ubeh. 115 Carl Størmer.)

history of the international congresses and of the Fields Medals.)

The astrophysicist Carl Størmer, president of the organizing committee, was elected president of the congress at the proposal of Rudolf Fueter, president of the 1932 congress. As in 1932, there was a poem recited at the congress. In his opening address, Størmer included a poem written by the Norwegian poet Bjørnstjerne Bjørnson (who had won the Nobel Prize for Literature in 1903) for the commemoration of the centenary of Abel's birth in 1902:

> Impassible as time
> the science of numbers is.
> Its combinations are
> in an eternal aurora
> purer than snow,
> subtler than air,

yet stronger than the world,
which without scales, weigh,
and without beams, illuminate.

(Any coincidence with the original is pure chance, since the poem was first translated from Norwegian into French and then from French into English).

After the inaugural session, a group photograph of all the congress participants was taken near the Aula in front of a statue of the historian Peter Andreas Munch (uncle of the expressionist painter) (see page 103). Almost in the center of the photograph, we can easily identify Gaston Julia with the mask over his face.

That afternoon, the congress members were received by King Haakon and Queen Maud at tea in the Royal Castle. The scenery was described by Waldo Dunnington: "The castle is surrounded by a

Group photograph of the 1936 congress. (Courtesy of the National Library of Oslo NBO Ubeh. 115 Carl Størmer.)

stately park where are a portrait bust of Camilia Collet, the pioneer Norwegian feminist, and a monument of Abel, both by the great Norwegian sculptor Gustav Vigeland."

There were a large number of plenary lectures, and their topics were diverse. There were 19 forty-five minute lectures, which, in chronological order, were the following:

- "Programme for the Quantitative Discussion of Electron Orbits in the Field of a Magnetic Dipole, with Application to Cosmic Rays and Kindred Phenomena," by Carl Størmer from Oslo;

- *"Die Theorie der regulären Funktionen einer Quaternionenvariablen,"* by Rudolf Fueter from Zurich;

- *"Quelques aperçus sur le rôle de la théorie des groupes de Lie dans l'évolution de la géométrie moderne,"* by Élie Cartan from Paris;

- *"Analytische Theorie der quadratischen Formen,"* by Carl L. Siegel from Frankfurt am Main;

- "Spinors and Projective Geometry," by Oswald Veblen from Princeton;

- *"Einige Methoden und Ergebnisse aus der Topologie der Flächenabbildungen,"* by Jakob Nielsen from Copenhagen;

- *"Neuere Fortschritte in der Theorie der elliptischen Modulfunktionen,"* by Erich Hecke from Hamburg;

- *"Über griechische Mathematik und ihr Verhältnis zur Vorgriechischen,"* by Otto Neugebauer from Copenhagen;
- *"Probleme der geometrischen Optik,"* by C. W. Oseen from Stockholm;
- "New Lines in Hydrodynamics," by Vilhelm Bjerknes from Oslo;
- *"Über die Riemannsche Vermutung in Funktionenkörpern,"* by Helmut Hasse from Göttingen;
- "The Foundations of Quantum Mechanics," by George David Birkhoff from Cambridge, MA;
- "Minkowski's Theorems and Hypotheses on Linear Forms," by Louis J. Mordell from Manchester;
- *"Geometrie der Riemannschen Flächen,"* by Lars V. Ahlfors from Helsingfors;
- *"Diophantische Approximationen,"* by Jan G. van der Corput from Groningen;
- *"Die Theorie der Operationen und ihre Bedeutung für die Analysis,"* by Stefan Banach from Lwów;
- *"Mélanges mathématiques,"* by Maurice Fréchet from Paris;
- "Gap Theorems," by Norbert Wiener from Cambridge, MA;
- "On the Decomposition Theorems of Algebra," by Øystein Ore from Yale.

For the first time, the proceedings of the congress included a report on the work of the recipients of the Fields Medals: *"Bericht über die Verleihüng der Fieldsmedallien,"* by Constantin Carathéodory from Munich. Carathéodory first explained that Ahlfors was "one of the most brilliant representatives of the famous Finnish school of function theory, founded by Ernst Lindelöf," and praised his work on the Nevanlinna theory for meromorphic functions. Regarding Douglas' work, Carathéodory referred to his "absolutely original method, which uses very few elements of the existing theory which allowed him to obtain far reaching and unexpected consequences" for the solution of Plateau's problem.

Afternoons during the congress were devoted to the sections. These were very similar to those in Zurich in 1932, except for small changes in some sections:

- Section III: Geometry and Topology,
- Section IV: Calculus of Probability, Mathematical Statistics, Actuarial Mathematics, and Econometrics,
- Section V: Mathematical Physics, Astronomy, and Geophysics,
- Section VI: Rational and Applied Mechanics,
- Section VII: Logic, Philosophy, and History,

For the first time, topology appeared, accompanying geometry; Section IV on statistics and calculus of probability included econometrics; the applied Sections V and VI reflected the interests of the Scandinavian scientific communities; and in Section VII, along with philosophy, logic was included. All these additions reflected the development that corresponding mathematical areas were experiencing.

A sign of new developments in international collaboration was the information presented to the congress by Solomon Lefschetz from Princeton, about an international conference on topology to be organized in Warsaw in 1939 by Sierpiński, "if circumstances made it at all possible," a phrase that probably, at that moment, was meant more as a formality than as a meaningful observation.

The historical reminiscences had a special Norwegian flavor in this congress. On Wednesday, July 15, a bust of Sophus Lie by the Norwegian sculptor Dyre Vaa was unveiled at the University of Oslo (see page 105). The bust was a gift to the university from a committee of private donors. A congratulating telegram was sent to Friedrich Engel from Giessen, who had been a pupil of Lie and had prepared the edition of

In Section VIII, the project of publishing a large collection of letters of Jean (Johann) Bernoulli was presented as part of a donation from a Basel industrialist to create the Bernoulli Foundation. As before, this section hosted the sessions of the International Commission on Mathematical Instruction, which were not chaired by Hadamard because he was on a trip to China. The commission had maintained its activity since 1932. It was invited by the congress to continue with its duties.

There were 487 mathematicians attending from 36 countries. The low attendance might have been caused by concerns about political instability and the economic crisis of the time. The congress was very Anglo-Saxon with respect to its participation. The national groups from the U.S.A. and the U.K. were the largest, with 86 and 48 participants, respectively. These groups constituted more than a quarter of the whole congress. The participation of women was proportionally larger than in 1932. The number was almost the same, 35, but general attendance was smaller.

Sophus Lie (1842–1899). (From *The Mathematician Sophus Lie* by Arild Stubhaug, Springer 2002.)

Lie's collected works. The plenary lecture of Cartan included his personal memories of Lie.

A bust of Sophus Lie was unveiled at the 1936 congress. (From *The Mathematician Sophus Lie* by Arild Stubhaug, Springer 2002.)

RÉPARTITION DES MEMBRES DU CONGRÈS PAR DIFFÉRENTS PAYS

	Participants;	Personnes adhérentes;		Participants;	Personnes adhérentes;
			Transporté	276	107
L'Union Sud-Africaine	2	1	Hongrie	5	1
Algérie	1	—	Inde	3	—
Allemagne	35	2	Iran	2	1
Australie	1	—	Irlande	6	3
Autriche	10	2	Islande	1	—
Belgique	9	3	Italie	5	2
Bulgarie	2	—	Japon	4	—
Canada	7	6	Lettonie	3	1
Chine	1	—	Norvège	59	25
Danemark	22	5	Pays-Bas	15	4
Egypte	3	—	Palestine	5	1
Espagne	8	—	Pologne	25	4
Esthonie	1	—	Roumanie	9	6
Etats-Unis	86	56	Suède	26	10
Finlande	8	1	Suisse	20	11
France	28	13	Tchécoslovaquie	10	3
Grande-Bretagne	48	18	U.R.S.S.	11	—
Grèce	4	—	Yougoslavie	2	3
A transporter	276	107	Total	487	182

Distribution of members of the 1936 congress according to nationality. (From the proceedings of the 1936 ICM, A.W. Brogers Boktrykkeri A/S 1937.)

Corresponding to the decrease in the number of participants, there was also a decrease in the number of communications presented, 205. Among these, we find that of Fritz Noether from Tomsk, brother of Emmy Noether, who had just died the previous year. His lecture was *"Über elektrische Drahtwellen."*

The situation of Italy and the Soviet Union was of special interest. From Italy, only five participants were listed, but most of them did not attend. We have already noted the absence of Severi, who had chaired the Fields Commission. The Italian government had decided not to allow attendance at the congress as a protest against Norway's backing of sanctions against Italy for its invasion of Abyssinia (Ethiopia). In the case of the Soviet Union, eleven participants were listed, but it seems that none of them attended, not even the ones invited to deliver plenary lectures. Alexandr Gelfond, who had planned to lecture on *"Théorie des nombres transcendants,"* and Alexandr Khintchine, on *"Hauptzüge der modernen Wahrscheinlichkeitstheorie,"* did not attend.

The closing of the congress took place in the Aula. The session began with the approval to send telegrams to Hilbert, Picard, and Volterra (this was another sign of reconciliation, since Picard and Volterra had been strongly involved in the exclusion policy of the International Mathematical Union; in the case of Volterra, it was also a sign of support, since he had been expelled from the university for not swearing fidelity to Mussolini's regime). Important decisions were taken related to the future of the international congresses.

First, the Fields Committee in charge of deciding on the two medalists for the next congress was chosen. Its members were Godfrey H. Hardy from Cambridge as president, Pavel Alexandrov from Moscow, Erich Hecke from Hamburg, Gaston Julia from Paris, and Tullio Levi-Civita from Rome. A month after the congress, Hardy resigned and was replaced by Solomon Lefschetz.

Regarding the activity of the commission for studying the future of the International Mathematical Union created during the Zurich 1932 congress, Gaston Julia, acting as president in absence of Severi, reported that after several meetings and deliberations, no unanimous conclusion had been reached. The congress accepted that situation; this was the formal end to the union.

Then, the venue for the next congress had to be decided. Luther P. Eisenhart from Princeton presented the invitation to hold the 1940 congress in

> the United States of America, the place of the meeting to be determined later by the society. This invitation is presented by the official delegates of the Society in accordance with action taken by the Council of the Society.

The last sentence was very important because it was precisely Eisenhart who, 16 years before at the Strasbourg 1920 congress, had issued the invitation to hold the 1924 congress in New York without the consent of his society, something that caused the congress to be transferred to Toronto. The invitation was warmly approved. It was commented that the possible venue for the congress was either New York or a smaller city on the Atlantic seaboard.

Let us just record the simple finances of the congress: 45,000 crowns, of which 10,000 came from the Ministry of Culture and Public Instruction and 35,000 from the associations of Norwegian banks and insurance companies.

Now that the International Mathematical Union had been definitively buried, the congress was again in total control of its meetings and concerns. The prospects for future international scientific collaboration for mathematicians again seemed sound and clear. The president of the congress, Carl Størmer, expressed this sentiment, along the lines of the words of Adolf Hurwitz at the first congress in 1897:

> Possibly the most important achievement of a congress such as this one results not from communications

and lectures but from informal conversations between mathematicians from different parts of the world. The direct intercourse of ideas in the form of conversation has an importance which, without leaving any trace in the proceedings of the congress, will however be manifest in the mathematical literature in the coming years.

After having followed the path of the international congresses from 1920 to 1936, we can appreciate the dramatic memories of the war described by Gaston Julia at the reception in the Bristol Hotel in Oslo in 1936. Those words represent both his personal struggle and the struggle of the mathematical community to survive and to overcome the effects of the atrocities of the war. The mathematical community had been able, in the long run, to maintain its will to collaborate internationally, and the ICM was the symbol of that spirit.

Meanwhile, outside of the Aula, the world was steadily moving towards another catastrophe. Indeed, the closing ceremony took place on Saturday, July 18, the day in which a military *coup d'état* started the Spanish Civil War, a terrifying rehearsal of World War II.

The German Delegation to the Oslo 1936 Congress had their own ceremony in memory of Niels Henrik Abel where they deposited a wreath at Abel's monument. (Courtesy of Nils Voje Johansen.)

Interlude

AWARDS OF THE ICM

SCIENCE IS A CHALLENGE, and mathematics is so to an even greater degree, because mathematics is the most problem-solving oriented of all sciences. The end of the Middle Ages witnessed a new era for science with the recovery of the scientific tradition of Greece. One of the many innovations was that science became a more collective activity. This is seen, for example, in the Renaissance contests associated with the development of algebra in Italy during the sixteenth and seventeenth centuries, and in the competitions for the solving of analytical problems during the Scientific Revolution in the seventeenth and early eighteenth centuries. Then, learned societies arose and competitions began to be organized into a system of prizes and awards. The Académie des Sciences of Paris created the Grand Prix in 1721, and the Royal Society of London created the Copley Medal in 1731. The next century witnessed the rise of the mathematical societies and with them other more mathematically oriented awards, such as the Steiner Prize of the Berlin Academy or the De Morgan Medal of the London Mathematical Society. All these awards had their limitations. Some were focused on a certain topic, or the nationality of the solvers was restricted, or they had publication constraints (regarding time or place), or they were shared with other sciences. Thus, with the creation of the Fields Medal in the Zurich 1932 congress and its first awarding in the Oslo 1936 congress, mathematics acquired a truly international award "open to the whole world" (phrase from the Fields memorandum).

The obverse side of the Fields Medal, featuring Archimedes' head. (From the author's personal files.)

However, awards have not always received general acceptance. When John Ligton Singe presented Fields' project of an international award in mathematics at the Zurich 1932 congress, he reported that there was "a little opposition from some who disapproved of such prizes."

In any case, awards do play a role in the system of science. The driving force for scientific work is (or at least has been up to recent times) a mixture of personal pride and self satisfaction, together with the appreciation from the scientific community. Here is where prizes enter. However, there can be side effects, as Alain Connes, Fields medalist in the Warsaw 1982 congress, explains:

> The utility of the awards is a delicate issue. In principle, the Fields Medal is an encouragement for research. But, if it attracts too much publicity, it can have negative effects. Jean-Paul Sartre, when he received the Nobel Prize in Literature, said that receiving a prize is like the kiss of death because if it is not well assimilated, it can annihilate people. Unfortunately, in some cases this is true.

THE FIELDS MEDAL

It is proposed to found two gold medals to be awarded at successive International Mathematical Congress for outstanding achievement in mathematics. Because of the multiplicity of the branches of mathematics and taking into account the fact that the interval between such Congresses is four years it is felt that at least two medals should be available. The awards would be open to the whole world and would be made by an International Committee.

This is the opening of the memorandum entitled "International Medals for Outstanding Discoveries in Mathematics," which is the founding document of the award known as the Fields Medal. Nowadays, it is the most distinguished international award in mathematics (even after the creation in 2002 of the Abel Prize by the Norwegian government). It is awarded by the International Mathematical Union every four years on the occasion of the International Congress of Mathematicians in recognition of "outstanding achievements in mathematics." At the same time, it is intended to be an encouragement for further achievements on the part of the recipients.

The Fields medalists are chosen by a Fields Medal Committee appointed by the union and normally chaired by its president. The committee is asked to select two to four medal recipients, preferably four. The committee is advised that, in the choosing, the diversity of mathematical fields should be taken into account. One of the peculiarities of the Fields Medal is that candidates must be less than 40 years old. In fact, the rule is much more precise; the candidate's 40th birthday must not have occurred before January 1 of the year of the congress in which the medals are awarded.

The presentation of the medals constitutes the highlight of the opening ceremony of the congress, when the secret, securely kept until that moment, is revealed: the names of the awardees (and also the names of the rest of the members of the Fields Medal Committee).

The alert reader should have noticed that the Union Mathématique Internationale was dissolved in the 1930s. The mystery of its reappearance and its current role in the international mathematical collaboration is disclosed in Part III.

The award consists of a medal and a small monetary prize. The medal is struck in gold by the Royal Canadian Mint and is 64 millimeters in diameter. The obverse side of the medal shows the head of Archimedes, the great mathematician (scientist) of antiquity. (He stands together with Newton and Gauss as probably the greatest of all time.) Archimedes is facing right. Around the medal is the inscription

$$\text{APXIMH}\Delta\text{O}\Upsilon\Sigma$$

which means "of Archimedes." On the left is the inscription

RTM
MCNXXXIII

which are the initials of Robert Tait McKenzie, the Canadian sculptor who designed and carved the medal, and in Roman numerals the year 1933, when the medal was created. There is a typing error in the date: the third letter should be M instead of N. Encircling theses three elements is the Latin inscription

TRANSIRE SVVM PECTVS MVNDOQVE POTIRI

which means "To transcend oneself and master the world" and which is adapted from verses of the poem "Astronomicon" of the Roman writer Manilius (book IV, verse 392).

The reverse side of the medal has the inscription

CONGREGATI
EX TOTO ORBE
MATHEMATICI
OB SCRIPTA INSIGNIA
TRIBVERE

which may be translated "Mathematicians, having congregated from all over the world, awarded [this medal] because of noteworthy writings." Behind the

The reverse side of the Fields Medal. In the background, the drawing on Archimedes' tomb of a sphere inscribed in a cylinder. (From the author's personal files.)

inscription there is a laurel branch, and in the background there is a sphere inscribed in a cylinder, following Plutarch's and Cicero's account of the drawing that Archimedes requested to be engraved on his tomb. The inscriptions were composed by G. Norwood from the University of Toronto. The name of the medalist is engraved on the rim of the medal.

The exceptional portrait of Archimedes shown on the medal is said to have been inspired by pictures in a fine collection of over 30 images of Archimedes collected by D. E. Smith at the University of Columbia. We are lucky to read of the sculptor's intention:

I feel a certain amount of complacency in having at last given to the mathematical world a version of Archimedes which is not decrepit, bald-headed, and myopic, but which has the fine presence and assured bearing of the man who defied the power of Rome.

How was this award created? The best way to discover the answer and to appreciate the peculiarities of the process is to follow the minutes of the organizing committee of the 1924 congress, which had John Charles Fields as chairman and John Lighton Synge as secretary, both from the University of Toronto. In the meeting of February 24, 1931, it was reported that, after meeting the expenses of the congress and the cost of printing the proceedings, there was a balance left of over 2700 Canadian dollars. Then,

It was resolved that the sum of $2500 should be set apart for two medals to be awarded in connection with successive International Mathematical Congresses through an international committee appointed for such purpose initially by the executive of the International Mathematical Congress, but later by the International Mathematical Union, the total cost of Medals to be around $400.

The next meeting of the committee took place one year later, on January 12, 1932. There, among other decisions,

It was decided to allot a maximum amount of $600 for the design of the medal.

It was decided that the Chairman should see the Prime Minister of Canada to arrange if possible how permanence of capital and of interest of the fund might be assured.

The Chairman reported that the following bodies had expressed approval of the scheme for the presentation of the International medal: American Mathematical Society, Société Mathématique de France, Deutsche Mathematiker-Vereinigung, Société Mathématique Suisse, Circolo Matematico di Palermo.

This shows that the lobbying that Fields had done searching for support for his idea of the award had been successful. Attached to the minutes of the meeting is the memorandum we have cited above, where the award was outlined; it is signed by Fields and undated.

The next step was presenting the proposal of the award for final approval at the International Congress of Mathematicians, which was meeting in September 1932 in Zurich. Fields was going to do this himself, making use of his prestige as organizer of the 1924 congress and of his known support of international collaboration. However, he suffered a heart attack in May. Let us follow the story from the words of Synge:

> When he was near death, he sent for me to be present with his lawyer when he made his will. He could hardly speak and the lawyer had some difficulty in making out what he wanted to be done.

After some monetary bequests were made, Fields' last will was:

> To transfer and pay over the balance of the residue of my Estate to John Lighton Synge, ... and the person for the time being the Premier of the Dominion of Canada in trust ... for the purpose of providing out of income thereof prizes to be attached to the International Mathematical Congress and also Medals.

In early August, Fields succumbed to a cerebral hemorrhage. Thus, it was Synge who presented the proposal of the medal to the Zurich 1932 congress. As we saw when the congress was described, the proposal was accepted with thanks, and the first Fields Medal Committee came into existence.

The next meeting of the organizing committee of the 1924 congress was held one year later, on January 16, 1933. There, the acceptance of the award by the international congress was acknowledged, and the contents of Fields' will revealed. The amount that Fields added to the medal funds is said to have been 47,000 Canadian dollars! The minutes of the meeting tell us that:

> A letter from the Master of the Royal Canadian Mint was read. At present rates the cost, inclusive of striking and material for two medals, would be $414 for fine gold and $308 for 18 carat gold, if the size is 2-13/16 in. = 7.14 cm. It was resolved that the medals should be struck at the Mint.

> It was resolved to offer the commission of the design of the medals to Dr. R. Tait McKenzie of Philadelphia, the fee to be $1100 (Canadian Funds), this fee to be inclusive of the design and the cutting of the dies for two faces.

At the same meeting, the committee was informed that the Dominion government would not undertake the duties of a trustee, so it was resolved to approach the Board of Governors of the University of Toronto with a similar request. In its final meeting, a year later, on January 4, 1934, the committee was informed that the University of Toronto would act as trustee for the Medal Fund but not for the Prize Fund. It is somewhat surprising the little attention that the Fields Medal has received from the government of Canada over the years.

We have already seen how and to whom the first two medals were awarded in 1936. As we will soon see, the following two medals had to wait to be awarded until 1950. In each of the 1954, 1958, and 1962 congresses, two medals were awarded. In the 1966 congress in Moscow, there was a novelty concerning the medals. The Indian Sir Dorabji Tata Trust intended to institute two new medals, similar to the Fields Medals. At the last moment, Indian authorities did not allow the money transfer, and the project was canceled. In any case, that year, due to an anonymous donor (whose name is still unknown), it was possible to award four Fields Medals. This was explained by Georges de Rham, president of the International Mathematical Union and of the Fields Committee, at the presentation of the medals at the opening ceremony in Moscow: "In view of the vast development of mathematics during the last forty years, it appears that [the number of medals] could judiciously be increased to four."

However, the key issue was to find the funds necessary for the medals and the cash prizes. For the 1970 congress, it was possible to maintain the awarding of four medals due to the accumulated income in the trusts of the medals and the prizes. The 1974 congress took place again in Canada, in Vancouver. At this congress, only two medals were awarded (although up to four might have been awarded). The balance left over after the congress was transferred to the Fields Medal funds held by the University of Toronto. This secured the future of the medals, guaranteeing funds for four medals. It also allowed for an increase in the amount of the monetary prize, which until the 1978 congress was 1500 Canadian dollars. It was subsequently doubled in 1983, and again in 1986; since 1990, it has been 15,000 Canadian dollars.

Regarding the Fields Medal Committee in charge of awarding the medals, we have already seen how the Zurich 1932 congress appointed a committee for the medals to be awarded in 1936. At the Oslo 1936 congress, the corresponding committee for the next medals was appointed. However, the lapse of 14 years before the next congress meant that it was the organizing committee of the 1950 congress who appointed a new Fields Medal Committee for the award-

ing the medals at that same congress. This was the procedure used for the 1954 and 1958 congresses. For the Stockholm 1962 congress, however, the Fields Medal Committee was appointed jointly by the organizing committee of the congress and the International Mathematical Union. From 1966 on, only the union has been entrusted with appointing the Fields Medal Committee.

Let us go back and look at the memorandum that Fields prepared that was approved at the 1932 congress. A copy of the memorandum is kept at the Archives of the International Mathematical Union in Helsinki (see page 114). The guidelines that constitute the award are outlined: the special role that the presentation of the medals should have in the congresses; the usage of Latin and Greek in the medal because of the international character of the award; its importance for international scientific cooperation; the future role to be played by the International Mathematical Union.

A peculiar requirement of the memorandum was that

[I]n making the awards while it was in recognition of work already done it was at the same time intended to be an encouragement for further achievement on the part of the recipients.

This is the most original of the clauses of the award and distinguishes the Fields Medal from other scientific awards. It was interpreted by the first Fields Committee, in Oslo in 1936, as meaning that the award should be given to "two young mathematicians," following the view expressed by the 1932 congress. In 1950, it was remarked by Harald Bohr, president of the committee, that the instructions were that the medals be given to "two really young mathematicians, without exactly specifying, however, the notion of being 'young.'" In 1958, the president of the committee, Heinz Hopf, explained that the committee had agreed "to keep the tradition of awarding the medals to mathematicians of the younger generation." The definite settlement of the age

Copy of Fields' memorandum, in the IMU Archives in Helsinki. (Courtesy of the Archives of the International Mathematical Union at the University of Helsinki.)

issue came during the 1966 congress in Moscow, where the tradition of considering only young mathematicians was specified and put into writing. Again, it was de Rham, at the presentation of the medals in the opening ceremony in Moscow, who explained that "On the basis of this text [the memorandum], and following precedents, we confine our choice to candidates under forty."

This rule was later made more precise and has been applied strictly ever since. The rule is clear and its application simple. However, there are always cases that confront the rules. This occurred with the proof of Fermat's Last Theorem ($x^n + y^n = z^n$ has no nontrivial integer solutions for n larger than 2). In 1993, Andrew Wiles lectured on his proof of the theorem at the Isaac Newton Institute in Cambridge. However, when the result was written up for publication, it was discovered

that "one step in the argument was not complete." In the 1994 congress, held for the third time in Zurich, Andrew Wiles lectured on his proof of the theorem. At that moment, the problem in the proof was not solved. A few weeks afterwards, the proof was completed. Wiles was 41 years old. The theorem, which had resisted solution from the most preeminent mathematicians for more than 350 years, had finally been proved.

At the next congress, in Berlin in 1998, Wiles was 45, so he could not receive the Fields Medal. The International Mathematical Union decided to create "a commemorative silver plaque as a special tribute to Andrew Wiles on the occasion of his sensational achievement."

We conclude the discussion of the memorandum by noting two of the specifications made by Fields, which say a great deal about his meticulous and generous personality:

> In commenting on the work of the medalists it might be well to be conservative in one's statements, to avoid invidious comparisons explicit or implied.

> [T]he medals should be of a character as purely international and impersonal as possible. There should not be attached to them in any way the name of any country, institution or person.

However, the first time the medal was awarded in Oslo in 1936, despite the efforts of its creator, it was officially introduced as the Fields Medal.

John Charles Fields (1863–1932). (Courtesy of The Royal Society of London.)

Let us briefly look at "the man behind the medal." John Charles Fields was born in 1863 in Hamilton, Ontario, into a Canadian family of Scottish-Irish origins. He obtained his B.A. degree from the University of Toronto with a gold medal in mathematics. Since at that time he could not work towards a Ph.D. in mathematics in Canada, he went to Johns Hopkins University

for his graduate studies, receiving his Ph.D. in mathematics in 1887. He taught at North American universities until 1892 when he left to pursue further studies in Europe. There he remained for ten years—the first five in France, mainly in Paris, and the next five in Germany, mostly in Berlin, but he also visited Göttingen. In his notebooks there are notes from lectures by Fuchs, Frobenius, Hensel, Schwarz, and Weierstrass. In 1906, he published the treatise *Theory of Algebraic Functions of a Complex Variable*. This European experience marked him deeply. The extended trip was made possible by a modest personal income and his "simple living and abstemious habits."

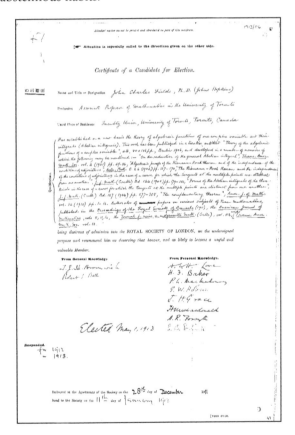

Certificate of election of J. C. Fields as member of the Royal Society of London, 1913. (Courtesy of The Royal Society of London.)

In 1902, he returned to the University of Toronto to a position as lecturer in mathematics. He remained there for the rest of his life, being finally appointed as Research Professor in 1923. He was engaged in many initiatives aimed at the promotion of research. His major task was the organization of the 1924 International Congress in Toronto, for which he displayed all his energies; it was after the long excursion organized for the congress to the western provinces of Canada that his health first deteriorated. He was a member of several scientific societies; in 1913, he was elected a member of the Royal Society of London.

During his stay in Europe, Fields developed an enduring friendship with Mittag-Leffler, something that seems to be the origin of Fields' interest in creating an international award in mathematics. He felt strongly about the lack of such an award and, probably through Mittag-Leffler, may have come to know the true reasons for mathematics being omitted from the legacy of Alfred Nobel (Gårding and Hörmander classified the existing different versions of this omission: the French-American version says that Mittag-Leffler had an affair with Nobel's wife; the Swedish version blames the rivalry between Mittag-Leffler and Nobel. Gårding and Hörmander conclude, however, that both versions are false: Nobel was not married, and he and Mittag-Leffler had almost no relationship).

We end this section by listing the 48 mathematicians who, up to the present date, have been awarded the Fields Medal. In addition to the year when the medal was awarded and the name of the awardee, we have decided to give the age of the recipient when he (no woman has yet received the award) received the medal (more precisely, his age on January 1 of the year of the congress when the medal was awarded) and his country of birth. This last information is both important and controversial. It is important because many mathematicians tend to count Fields medalists by their birthplace ("Finally Australia has a Fields Medal!"—that of Terence Tao in 2006; or "Japan has three Fields Medals!"—those of Kodaira in 1954, Hironaka in 1970, and Mori in 1990). But this is controversial because, for example, there are currently nonexisting countries (such as the Soviet Union) where many medalists were born, or a medalist could have been born in one country but raised, educated, and nationalized in a different one (such as W. Werner, who was born in Germany but has had French nationality since the age of nine).

Note that the list is that of medals awarded, not necessarily received. For example, as a protest against some of the Soviet Union's political actions, Alexander Grothendieck did not attend the Moscow 1966 congress, where he was awarded the medal; however, the medal was received on his behalf. During the Cold War, some awardees from the U.S.S.R. were not allowed to attend the congress where the medal was awarded (Sergei Novikov in 1970 and Gregori Margulis in 1978). But in all of these cases, the awardees accepted the medal and eventually received it. There has been only one case in which the Fields Medal has not been accepted. In 2006, Grigory Perelman explicitly refused it; in any case, the International Mathematical Union insisted on awarding him the medal.

It is beyond our scope to comment on the type of mathematics that has been rewarded with the Fields Medal in each case. However, the words of Heinz Hopf, president of the Fields Medal Committee in 1958, are enlightening in this regard:

> The great variety within mathematics is due not only to the multiplicity of branches of mathematics, but also to the diversity of the general tasks that face a mathematician in any branch. A task which is particularly fundamental is: to solve old problems; another, no less fundamental, is: to open the way to new developments.

Year	Name	Country of birth	Age
1936	Lars V. Ahlfors	Finland	29
	Jesse Douglas	U.S.A.	38
1950	Laurent Schwartz	France	34
	Atle Selberg	Norway	32
1954	Kunihiko Kodaira	Japan	38
	Jean-Pierre Serre	France	27
1958	Klaus F. Roth	Germany	32
	René Thom	France	34
1962	Lars Hörmander	Sweden	30
	John W. Milnor	U.S.A.	30
1966	Michael F. Atiyah	U.K.	36
	Paul J. Cohen	U.S.A.	31
	Alexander Grothendieck	Germany	37
	Stephen Smale	U.S.A.	35
1970	Alan Baker	U.K.	30
	Heisuke Hironaka	Japan	38
	Sergei Novikov	U.S.S.R.	31
	John G. Thompson	U.S.A.	37
1974	Enrico Bombieri	Italy	33
	David B. Mumford	U.K.	36
1978	Pierre René Deligne	Belgium	33
	Charles L. Fefferman	U.S.A.	28
	Gregori A. Margulis	U.S.S.R.	31
	Daniel G. Quillen	U.S.A.	37
1982	Alain Connes	France	34
	William P. Thurston	U.S.A.	35
	Shing-Tung Yau	China	32
1986	Simon K. Donaldson	U.K.	28
	Gerd Faltings	Germany	31
	Michael H. Freedman	U.S.A.	34
1990	Vladimir Drinfeld	U.S.S.R.	35
	Vaughan F. R. Jones	New Zealand	37
	Shigefumi Mori	Japan	38
	Edward Witten	U.S.A.	38
1994	Jean Bourgain	Belgium	39
	Pierre-Louis Lions	France	37
	Jean-Christophe Yoccoz	France	36
	Efim Zelmanov	U.S.S.R.	38
1998	Richard E. Borcherds	South Africa	38
	W. Timothy Gowers	U.K.	34
	Maxim Kontsevich	U.S.S.R.	33
	Curtis T. McMullen	U.S.A.	39
2002	Laurent Lafforgue	France	35
	Vladimir Voevodsky	U.S.S.R.	35
2006	Andrei Okounkov	U.S.S.R.	36
	Grigory Perelman	U.S.S.R.	39
	Terence Tao	Australia	30
	Wendelin Werner	Germany	37

The 48 mathematicians who have been awarded the Fields Medal.

Lars Hörmander and John Milnor, 1962 medalists. (Courtesy of the Center for History of Science of the Royal Swedish Academy of Sciences.)

Those such as Michael Monastyrsky, who have studied this issue in detail (see his book *Modern Mathematics in the Light of the Fields Medals*), consider that in the awarding of the Fields Medal there has been a balance between these two poles.

THE NEVANLINNA PRIZE

Rolf Herman Nevanlinna was born in 1895 in Joensuu, a city in the eastern part of the Grand Duchy of Finland, at that time part of the Russian Empire. Many members of his family—his father, uncle, and older brother—had shown a talent for mathematics. He studied at the University of Helsinki under the guidance of Ernst Lindelöf, defending his thesis in 1919. He entered the University of Helsinki in 1922 and was appointed full professor in 1926. He was rector of the university in the war years, from 1941 until 1944. In 1946, he accepted a position at the University of Zurich and was also a member of the Academy of Finland until his retirement in 1963. Scientifically, he is known for the so-called Nevanlinna theory, originated in his work in complex analysis, and for his deep involvement in the development of the theory of harmonic measures.

Nevanlinna had a longstanding relation with the ICM and the International Mathematical Union. Following the path of his teacher Lindelöf, who had attended all congresses from 1897 to 1912, Nevanlinna began attending the congresses in 1928. In 1936, he had the pleasure of seeing his student (jointly with Lindelöf), Lars V. Ahlfors, receive the first Fields Medal. From 1959 to 1962, he was president of the International Mathematical Union; in 1962, he chaired the Fields Medal Committee and presided over the 1962 and (honorarily) the 1978 congresses. He died in Helsinki in 1980.

Rolf Nevanlinna (1895–1980). (Courtesy of Olli Letho.)

In 1981, when Lennart Carleson was president of the International Mathematical Union, the union decided to create (to a great extent due to Carleson's effort) a prize similar to the Fields Medal, with the goal that mathematics not lose contact with theoretical computer science. In the search for funding for the award, an offer came from the University of Helsinki. Since the natural choices for the name of the prize, such as Archimedes or von Neumann, were already in use, it was suggested to name it after Nevanlinna, honoring his efforts in the 1950s to introduce computers into the Finnish academic world.

The official name of the award is the Rolf Nevanlinna Prize in Mathematical Aspects of Information Sciences. It is awarded for outstanding contributions in the field, including

1. all mathematical aspects of computer science, such as complexity theory, logic of programming languages, analysis of algorithms, cryptography, computer vision, pattern recognition, information processing, and modeling of intelligence;

2. scientific computing and numerical analysis, computational aspects of optimization and control theory, and computer algebra.

The obverse side of the Nevanlinna Medal. (From the author's personal files.)

The prize consists of a gold medal and a monetary prize. One prize is awarded every four years at each international congress. The rest of the regulations are similar to those of the Fields Medal.

On the obverse side, the medal presents the head of Nevanlinna and the inscription RH 83, which stands for Raimo Heino, the Finnish sculptor who designed the medal (and, more recently, the Finnish two euro coin), and the year 1983, when the first medal was minted. The reverse side shows the seventeenth-century seal of the University of Helsinki, and the word Helsinki written in coded form. The name of the prize winner is engraved on the rim of the medal.

The reverse side of the Nevanlinna Medal. (From the author's personal files.)

The first medal was awarded in 1982, although it was presented to Robert Tarjan, the awardee, in 1983, for reasons that will be explained in Part IV.

The seven prize winners are listed on page 120. (The information displayed is similar to that of the Fields medalists.)

Year	Name	Country of birth	Age
1982	Robert Tarjan	U.S.A.	35
1986	Leslie Valiant	U.K.	37
1990	Alexander A. Razborov	U.S.S.R.	31
1994	Avi Wigderson	U.S.A.	38
1998	Peter W. Shor	U.S.A.	39
2002	Madhu Sudan	India	35
2006	Jon Kleinberg	U.S.A.	35

The seven mathematicians who have been awarded the Nevanlinna Prize.

THE GAUSS PRIZE

Recall that the Heidelberg 1904 congress ended with a financial surplus, which was graciously donated to the organization of the next congress, that of Rome in 1908. The 1998 congress was held again in Germany, this time in Berlin, and it was reported that there was also a surplus. In this case, the decision of the Deutsche Mathematiker-Vereinigung was to use the surplus to create a new international award in mathematics to be awarded jointly with the International Mathematical Union. This time the focus was placed on the influence of mathematics in other scientific disciplines.

The official press release reveals the aims of the award:

> Mathematics is an important and ancient discipline. However, it seems that only the experts know that mathematics is a driving force behind many modern technologies. The Gauss Prize has been created to help the rest of the world realize this fundamental fact. The prize is to honor scientists whose mathematical research has had an impact outside mathematics—either in technology, in business, or simply in people's everyday lives.

The announcement was made on April 30, 2002, on the occasion of the 225th anniversary of Gauss' birth.

Naming this award after Gauss was more than proper. Carl Friedrich Gauss was born in 1777 in Brunswick and died in 1855 in Göttingen. He was one of the greatest mathematicians of all times. Not in vain was he known by his contemporaries as *Princeps mathematicorum*; that is, among the mathematicians, he was the first. Mathematicians admire his *Disquisitiones arithmeticae*, still a masterpiece of scientific research. But for the layman, or even for the general scientist, of his time, he became a celebrity when he was able to determine the orbit of the newly discovered planetoid Ceres, based on very limited observational data. It is said that Laplace exclaimed:

Carl Friedrich Gauss (1777–1855). Portrait painted in 1840 by the Danish artist Christian Albrecht Jensen. (Courtesy of the Berlin Brandenburgische Akademie der Wissenschaften.)

"The duke of Brunswick has discovered more in his land than a planet: a super-terrestrial mind in a human body." (The Duke of Brunswick was Gauss's benefactor.) Lagrange wrote to Gauss, after praising the *Disquisitiones*, that "Your work on planets will in addition have the merit of the importance of the topic."

Gauss left his stamp on many practical activities, from the heliotrope, designed for the surveying of the State of Hannover, to the normal probability distribution, represented by the bell-shaped curve, and in the theory of electric and terrestrial magnetism. Gauss masterfully combined the abstract essence of pure mathematics with the concrete work of practical applications.

The obverse side of the Gauss Medal: a portrait of Gauss dissolved into a barcode of lines. (From the author's personal files.)

The official name of the prize is the Carl Friedrich Gauss Prize for Applications of Mathematics. It is to be awarded for outstanding

- mathematical contributions that have found significant practical applications outside of mathematics, or
- achievements that made the application of mathematical methods to areas outside of mathematics

possible in an innovative way, e.g., via new modeling techniques or the design and implementation of algorithms.

One prize is awarded every four years at the international congress, in the same manner as the other IMU awards. The only difference being that, since the applicability for practice of mathematical results may only be realized after a long interval of time, there is no age limit restriction for the prize winner. The prize consist of a gold medal and a monetary award (of 10,000 euros in 2006).

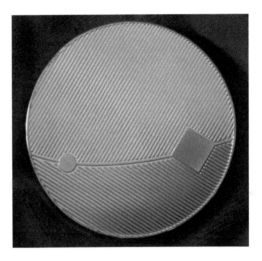

The reverse side of the Gauss Medal: symbolic representation of the discovery of Ceres' orbit. (From the author's personal files.)

We are lucky to have an explanation of the design of the Gauss Medal by Jan Arnold, the artist who conceived it:

Soon after Giuseppe Piazzi discovered the celestial body Ceres on January 1, 1801, Ceres disappeared from view, and there were no reliable techniques available to predict its orbit from Piazzi's limited observational data. Introducing a revolutionary new idea, the now well-known least squares method, Gauss was able to calculate Ceres' orbit in a very precise way, and

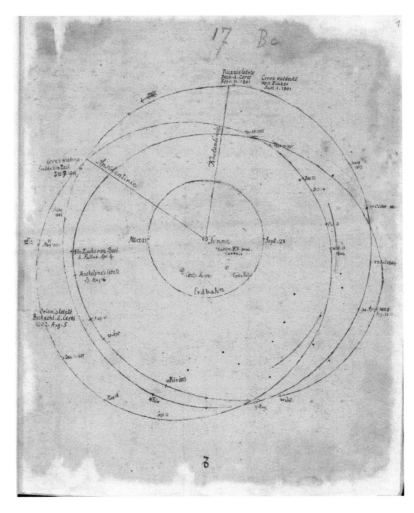

The orbits of Ceres and Pallas, by Gauss. (Courtesy of the Niedersächsische Staats- und Universitätsbibliothek Göttingen. Cod. Ms. Gauß Handbuch 4, Bl. 1v.)

in December 1801, Ceres was rediscovered by the astronomer Zack very close to the predicted position.

This impressive example illustrating the power of the applications of mathematics provided the general idea for the design of this medal.

Dissolved into a linear pattern, Gauss' effigy is incomplete. It is the viewer's eye which completes the barcode of lines and transforms it into the portrait of Gauss.

A similar pattern, accomplished by horizontal lines, is one of the features on the back of the medal. This grid is crossed by a curve. The disk and the square, two elements connected by the curve, symbolize both the least squares method and the discovery of Ceres' orbit.

The mathematical language has been reduced to its most fundamental elements, such as point, line and curve. Moreover, these elements represent natural processes. The imagery of the medal is a synthesis of nature's and mathematics' sign language.

At the 2006 international congress held in Madrid, the Gauss Prize was awarded for the first time to the Japanese mathematician Kiyoshi Itô, born in 1915, for his development of stochastic analysis, which has provided a tool for dealing with highly unpredictable phenomena, such as prices on financial markets or sizes of populations of living organisms.

Today, we see Gauss as a champion of applicability of mathematics, but, paradoxically, Gauss received some criticism from his contemporaries for the lack of usefulness of his work ... in astronomy! The astronomer Zack defended Gauss publicly by publishing in his astronomical journal, *Monatliche Correspondenz*, the following epigram by the French astronomer Lalande:

Résultat d'un Calcul mathématico-politique et moral,
par le Citoyen La Lande, Doyen des Astronomes

Il y a mille millions d'habitants sur la surface de la terre.
Sur ces mille millions de têtes
Que de méchants, de foux, de bêtes,
Mais nous ne pouvons les guérir,
Il faut les plaindre, et les servir.

This translates to

Outcome of a mathematic-political and moral calculation,
by citizen La Lande, Dean of astronomers

There are thousands of millions of inhabitants on the surface of the Earth.
Of those thousands of millions of heads
How many villains, fools, and idiots,
Since we cannot cure them,
We must commiserate and serve them.

PART III
THE GOLDEN ERA

In view of the distressful world situation, the International Congress of Mathematicians which was going to be held at Cambridge, Massachusetts, in September, 1940, was postponed. An Emergency Executive Committee, consisting of Professors G. D. Birkhoff, W. C. Graustein, Einar Hille, M. H. Ingraham, J. R. Line, Marston Morse, R. G. D. Richardson, and M. H. Stone, was appointed to act during the interim.

T HIS SHORT NOTE, which appeared in the November 1939 issue of the *Bulletin of the American Mathematical Society*, certified what had been obvious since September 1, 1939, when the invasion of Poland by Hitler's Third Reich caused the outbreak of World War II. As had already been announced by the Society the year before, the 1940 congress was fully arranged by that time. It was going to take place from September 4 to 12, hosted by Harvard University and the Massachusetts Institute of Technology. The scientific program was designed, financial backing secured, housing and entertainment organized, even the fees, $10, were fixed.

After the war, plans for holding the congress were reissued. A crucial issue was who could participate in the congress. In this regard, there was a clear stand in the U.S. mathematical community that the congress should be open to mathematicians of all countries. Indeed:

> Those guiding the policies of the American Mathematical Society were insistent that there should be no international congress until such a time that the gathering could be truly international in the sense that mathematicians could be invited irrespective of national or geographic origins.

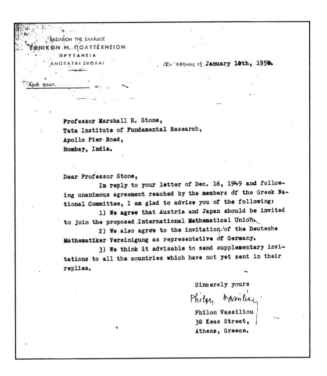

A survey was made: how to proceed after World War II. Answer of the Greek Mathematical Society. (Courtesy of the Archives of the International Mathematical Union at the University of Helsinki.)

This was in agreement with the viewpoint prevailing in all other scientific disciplines. These conditions having been fulfilled, the arrangements for the congress were resumed. At the same time, efforts were made for creating, not reestablishing, an International Mathematical Union; in this task, the actions of Marshall H. Stone were instrumental. The goals of the congress and of the union were different since the congress addressed the individual mathematician and the union addressed societies or countries. They were linked, though, because both required a positive attitude towards the atrocities of the war. Fortunately, memories of the disastrous aftermath of World War I for the ICM were still present and prevented further errors of the same sort.

By mid 1948, the announcement of the congress was released. It was similar to that of the 1940 congress, with one notable difference. To the official congress languages of English, French, German, and Italian, Russian was added.

Marshall H. Stone (1903–1989) led the refoundation of the International Mathematical Union after World War II. (Courtesy of the American Mathematical Society.)

Regarding the union, Stone was steadily reaching the objective. He began consulting with certain societies and individual mathematicians, and progressively he was able to obtain an agreement from a large number of countries. The most difficult issue was the participation of Germany and Japan. For the latter, Stone even contacted General Douglas McArthur, Supreme Commander of the Allied Powers in Japan. For the former, Stone waited to be contacted by the Deutsche Mathematiker-Vereinigung, which represented mathematicians from both East and West Germany. After consultation with the rest of the countries, there was a general approval for admitting Germany (see page 125). Thus, a Constitutive Convention was convened in New York City, at Columbia University in August 1950, just prior to the international congress.

These two actions led to the foundations of a new era of international cooperation that allowed the successful development of the ICM. The congresses gained acceptance, polished their procedures, and continued increasing their attendance. We have labeled this period "The Golden Era" because the foundations of the current international congresses were then laid out and because they displayed a classical and magnificent style. The congresses were

- Cambridge, MA, August 30–September 6, 1950;
- Amsterdam, August 2–9, 1954;
- Edinburgh, August 14–21, 1958;
- Stockholm, August 15–22, 1962.

CAMBRIDGE (MA) 1950

$\mathbf{T}$HE FIRST INTERNATIONAL CONGRESS after World War II was held at Harvard University under the auspices of the American Mathematical Society. At the opening plenary session, held on Wednesday afternoon, August 30, in Sanders Theatre (see page 128), there was an effort to resume the traditions of previous congresses. Indeed, Oswald Veblen, from the Institute for Advanced Study, was nominated for president of the congress at the proposal of Carl Størmer, who had presided over the previous congress, that of Oslo in 1936 (since Størmer did not attend the congress, his representative carried the proposal to Cambridge). After being unanimously elected, Veblen delivered an important presidential address, where he explained that for American mathematics, "the colonial period was ending" and that:

> We are approaching the end of another epoch. I mean the period during which North America has absorbed so many powerful mathematicians from all over the world that the indigenous traditions and tendencies of mathematical thought have been radically changed as well as enriched. These American gains have seemed to be at the cost of great losses to European mathematics. But there are so many signs of vitality in Europe that it is now possible to hope the losses will be only temporary while American gains will be permanent.

(We leave to the reader the consideration of whether or not Veblen's predictions were realized.)

Oswald Veblen (1880–1960), president of ICM 1950. (Courtesy of the American Mathematical Society.)

Another effort to link the congress to the past tradition of international congresses was the choice of honorary presidents, three for this congress: Guido Castelnuovo from the Accademia Nazionale dei Lincei, Jacques Hadamard from the Collège de France, and Charles de la Vallée Poussin from the Université de Louvain. All three were over 84 years old.

The Fields Medals were awarded for the second time, 14 years after the first medals awarded in Oslo. Because two of the five members of the committee appointed in Oslo had died, Hecke and Levi-Civita, a new committee was appointed using a different procedure (as we have seen in "Awards of the ICM");

Jacques Hadamard (1865–1963) at the 1950 congress, of which he was honorary president. (Courtesy of Jean-Pierre Kahane.)

Sanders Theatre at Harvard University. (Courtesy of the President and Fellows of Harvard College. Photo: Steve Rosenthal.)

it was the organizing committee of the congress who designated the Fields Medal committee, which in this case consisted of Ahlfors, Borsuk, Fréchet, Hodge, Kolmogorov, Kosambi, Morse, and Harald Bohr as chairman.

The medals were awarded to

* Laurent Schwartz from the Université de Nancy,
* Atle Selberg from the Institute for Advanced Study.

Bohr reported to the congress on the seminal work of Schwartz on the theory of distributions ("the new ideas in their purity and generality") and on Selberg's work on the Riemann zeta function and his elementary proof of the prime number theorem. Regarding Schwartz's work, Bohr recalled the words of Felix Klein, "great progress in our science is often obtained when new methods are applied to old problems," and with respect to Selberg's work, he quoted G. H. Hardy, who in 1921 had said that, "No elementary proof of the prime number theorem is known, and one may ask whether it is reasonable to expect one."

Intense scientific work started immediately after the opening ceremony with two parallel, one-hour addresses by mathematicians invited by the organizing committee. They were

* "On Null-Sets in Harmonic Analysis and Function Theory," by A. Beurling;
* *"Die n-dimensionalen Sphären und projektiven Räume in der Topologie,"* by H. Hopf;

and immediately after, there were another two:

* *"Sur les fonctions analytiques de variables complexes,"* by H. Cartan;
* "The Cultural Basis of Mathematics,'" by R. L. Wilder.

For the next two days, the morning started similarly with two parallel, one-hour addresses:

* "Laplace Operator on Manifolds," by S. Bochner;

"Rotating Universes in General Relativity," by K. Gödel;

and

"Recent Advances in Variational Theory in the Large," by M. Morse;

"The Calculation of an Eclipse of the Sun According to Theon of Alexandria," by A. Rome.

On Tuesday, September 5, there were six invited addresses delivered again in two parallel sessions:

"Basic Ideas of a General Theory of Statistical Decision Rules," by A. Wald;

"r-dimensional Integration in n-space," by H. Whitney;

"Topological Invariants of Algebraic Varieties," by W. V. D. Hodge;

"Differential Groups," by J. F. Ritt;

"Recent Progress in Geometry of Numbers," by H. Davenport;

"Distributions and Principal Applications," by L. Schwartz.

The reason for the somewhat stressful beginning and the intense program was the size and complexity of the scientific program of the congress. As in every other congress, there were invited plenary addresses, in this case 22, similar to the 19 plenary lectures of Oslo in 1936 and the 21 of Zurich in 1932. Also there were 374 ten-minute contributed papers presented in the scientific sections of the congress (there had been

Section number	Section name	Chairman	Invited speakers	Contributed papers
I	Algebra and Number Theory	H. A. Rademacher	Kloosterman Mahler Selberg	58
II	Analysis	G. C. Evans	Bergman Bohr Mandelbrojt Rademacher	127
III	Geometry and Topology	S. Eilenberg	Santaló Segre	58
IV	Probability and Statistics, Actuarial Science, Economics	J. L. Doob	Bose Lévy Roy	27
V	Mathematical Physics and Applied Mathematics	R. Courant	Darwin Lewy Rellich	74
VI	Logic and Philosophy	A. Tarski	Kleene Robinson Skolem Tarski	16
VII	History and Education	C. V. Newsom	Pólya	15

List of sections of the 1950 congress.

247 in 1932 and 205 in 1936). But there was a new feature in the sections. The chairman of each section was given the privilege of inviting no more than three persons to deliver thirty-minute addresses in the sections.

The topics of the sections were in the spirit of those of the last congresses. See the list on page 129.

George Pólya's lecture, entitled "On Plausible Reasoning," was on his ideas about the heuristics of mathematical research. (Pólya's famous book *How to Solve It* had been published just a few years earlier, in 1945.)

George Pólya's lecture in the 1950 congress. (From the proceedings of the 1950 ICM, American Mathematical Society 1952.)

For the first time, in the late 1930s, there had been conferences focused on a particular field of mathematics. They were very successful. One was a conference on topology held in Moscow, and another was a conference on probability held in Zurich. With this model, the organizing committee decided to hold, within the international congress, four conferences, each one devoted to a field where "vigorous advances have been made or are in progress," having a "well coordinated program of formal lectures and open informal discussion." Eight of the invited addresses corresponded to lectures in these conferences.

There were four of these conferences:

* Conference in Algebra,

  Chairman: A. A. Albert,

  Stated addresses:

  – "Power-Associative Algebras," by A. A. Albert;
  – "Number Theory and Algebraic Geometry," by A. Weil.
  – "The Fundamental Ideas of Abstract Algebraic Geometry," by O. Zariski;

* Conference in Applied Mathematics,

  Chairman: J. von Neumann,

  Stated addresses:

  – "Shock Interaction and Its Mathematical Aspects," by J. von Neumann;
  – "Comprehensive View of Prediction Theory," by N. Wiener;

* Conference in Analysis,

  Chairman: M. Morse,

  Stated address:

  – "Ergodic Theory," by S. Kakutani;

Commander Howard Aiken, designer of the Harvard Computing Machine; Grace Hopper; and the Harvard Mark I computing machine. (Courtesy of Harvard University Archives, Call # UAV 605.270.1.2p, U-822, U-823, U-824.)

Conference in Topology,

Chairman: H. Whitney,

Stated addresses:

- "Differential Geometry of Fiber Bundles," by S.-S. Chern;

- "Homology and Homotopy," by W. Hurewicz.

An interesting complement to the scientific program was a lecture by Howard Aiken on computing machines. Aiken was the designer of the series of electromechanical devices known as the Harvard Mark computing machines. These machines were built with funding from IBM (and the help of Grace Hopper) in the 1940s

This intensive activity lasted seven full days; there were even some sessions that started at 8 p.m.!

Thankfully, other activities entertained the congress participants. On opening day, there was a reception at night at the Fogg Art Museum of Harvard University. On Friday afternoon, September 1, the mathematicians and their friends were the guests of Wellesley College for tea. On Saturday evening, there was the choice of an informal dance at Lowell House or a beer party at Memorial Hall. Sunday was a free day, and musical entertainments were offered

(which we will review in "Social Life at the ICM"). The congress banquet was held on Tuesday evening, in the Sever Quadrangle. On Wednesday evening, September 6, the mathematicians were guests of the Director and Board of Trustees of the Gardner Museum for a farewell party.

How was the congress financed? The largest donors were the Carnegie Corporation and the Rockefeller Foundation, together with the American mathematical societies (American Mathematical Society, Mathematical Association of America); the nearby universities and scientific institutions (Harvard University, Boston University, Yale University, Massachusetts Institute of Technology, the Institute for Advanced Study); the UNESCO, who funded the publishing of the proceedings; and companies of all kinds: insurance, chemical, automobile (Ford, General Motors), oil (Shell, Standard Oil), publishing (John Wiley, Van Nostrand), and also General Electric, Bell, Eastman Kodak, United Fruit, and many others. The fee for participants had a 50 percent increase from the one announced for the 1940 congress; it was now $15.

At the closing session, Marshall H. Stone reported on the Constitutive Conference of the International Mathematical Union, which had been held in New York City immediately preceding the congress. The statutes and by-laws of an international union had been discussed and adopted and would be submitted to "the proper scientific groups in the various national and geographical areas where there was significant mathematical activity." It was then decided that when ten countries had indicated their acceptance, the union would be declared in existence. (This happened on September 1951, the first ten countries being Austria, Denmark, France, Germany, Great Britain, Greece, Italy, Japan, the Netherlands, and Norway. By December, 1951, another five countries joined the union: Australia, Canada, Finland, Peru, and the U.S.A. The first General Assembly of the union was held in Rome in 1952.)

AMSTERDAM 1954

THE INVITATION ISSUED at the closing of the 1950 congress by Johannes van der Corput to hold the next congress in the Netherlands was unanimously accepted. Thus, Amsterdam hosted the 1954 congress, held September 2–9, under the auspices of the Wiskundig Genootschap, the Netherlands mathematical society.

A splendidly decorated podium, with the flags of all countries represented at the congress, was set at the Concertgebouw of Amsterdam for the opening session of the congress. At Oswald Veblen's proposal, Jan A. Schouten, who had chaired the organizing committee and was president of the Wiskundig Genootschap, was elected president of the congress.

The presiding table at the opening ceremony of the Amsterdam 1954 congress. (From the Archive of the Centrum voor Wiskunde en Informatica, Amsterdam.)

Schouten delivered his welcome speech in Dutch, English, French, German, Italian, Swedish, and Russian (in contrast to this, the organization had decided that there would be no official languages for the congress).

After a musical interlude consisting of a piano solo by the pianist Fania Chapiro (who played Chopin's *Impromptu* op. 29, *Nocturne* op. 48, and *Scherzo no. 2* op. 31), the awarding of the Fields Medals commenced. Hermann Weyl had chaired the Fields Medal Committee, whose other members were E. Bompiani, F. Bureau, H. Cartan, A. Ostrowski, Å. Pleijel, G. Szegö, and E. C. Titchmarsh.

Weyl took the stand and announced the names of the awardees:

- Kunihiko Kodaira from Princeton University,
- Jean-Pierre Serre from the Collège de France.

Hermann Weyl addressing Kodaira and Serre, the 1954 Fields medalists. (From the Archive of the Centrum voor Wiskunde en Informatica, Amsterdam.)

Next he presented the medals. There is a vivid photograph of that moment, when Weyl seems to be recalling for Kodaira and Serre the instructions from the founder of the award regarding the need for pursuing further achievements. Afterwards, Weyl explained their work in detail: Kodaira's "outstanding achievements ... on the theory of harmonic integrals and the numerous profound applications ... to algebraic varieties"; and Serre's contributions to the homotopy theory of spheres "with the wealth of its surprisingly numerical results."

Another musical interlude followed (with Debussy's *Suite pour le Piano*). As part of the efforts to advertise the International Mathematical Union and to enhance its role in the international congresses, before this ceremony ended, the congress was informed of the election of Heinz Hopf as president of the union for the period 1955–1958.

Queen Juliana of the Netherlands received 1954 congress delegates and the IMU officers. (From the Archive of the Centrum voor Wiskunde en Informatica, Amsterdam.)

That afternoon, Thursday, September 2, 1954, a photograph of all the congress participants was taken in front of the Concertgebouw (see page 135). The magnificent picture gives a true image of Europe in the 1950s. Unfortunately, this was the last of the group pictures of an international congress. That evening, a reception was hosted by the government of the Netherlands and the Amsterdam Municipality at the Rijksmuseum.

The 1954 ICM in full.

The scientific program of the congress maintained the same format introduced in 1950. There were 20 invited lectures (see the list below), which, as had occurred at Harvard, had to run in parallel sessions of two or even three lectures. The only two exceptions were the plenary lectures delivered at the opening by von Neumann and at the closing of the congress by A. Kolmogorov :

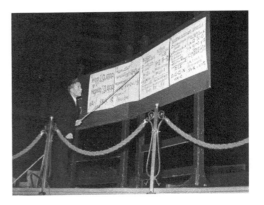

Andrey N. Kolmogorov lecturing at ICM 1954. (From the Archive of the Centrum voor Wiskunde en Informatica, Amsterdam.)

- "Unsolved Problems in Mathematics," by J. von Neumann;
- *"Théorie générale des systemes dynamiques et mécanique classique,"* by A. Kolmogorov.

John von Neumann lecturing at ICM 1954. (From the Archive of the Centrum voor Wiskunde en Informatica, Amsterdam.)

The other invited lectures, in chronological order of their delivery, were

- "Representations of Semi-Simple Lie Groups," by Harish-Chandra;
- "Eigenfunction Problems Arising from Differential Equations," by E. C. Titchmarsh;
- "On the Structure of Groups of Finite Order," by R. Brauer;
- "Recent Developments in Relaxation Techniques," by E. Stiefel;
- *"Le calcul différentiel dans les corps de caractéristique $p > 0$,"* by J. Dieudonné;

- "Some Aspects of the Theory of Almost Periodic Functions," by B. Jessen;
- *"Sur l'élimination de phénomènes paradoxaux en topologie générale,"* by K. Borsuk;
- "Current Problems in Mathematical Statistics," by J. Neymann;
- "Mathematical Problems Raised by the Flood Disaster 1953," by D. van Dantzig;
- "Mathematics and Metamathematics," by A. Tarski;
- *"Groupes de holonomie,"* by A. Lichnerowicz;
- "On Some Methods of Approximation in Fluid Mechanics," by S. Goldstein;
- *"Einige Fragen der Approximation von Funktionen durch Polynome,"* by S. M. Nikolsky;
- "Geometry upon an Algebraic Variety," by B. Segre;
- "Some Aspects of Functional Analysis and Algebra," by I. Gelfand;
- "Semi-group Theory and the Integration Problem of Diffusion Equations," by K. Yosida;
- "Abstract versus Classical Algebraic Geometry," by A. Weil;
- *"Aus der mengentheoretischen Topologie der letzten zwanzig Jahren,"* by P. Alexandrov.

The sections of the congress were similar to those of the 1950 congress, except for the last two sections; in Section VI, Foundations was the topic instead of Philosophy, which was shifted to Section VII. There were also 42 experts in the various branches of mathematics who were invited to give half-hour lectures related to the sections. An overall number of 496 fifteen-minute communications were presented in the sections.

With such a huge number of lectures of all sorts, most of the sessions were running simultaneously, and some of them had to be split into several parallel groups. This required an extraordinarily well-organized schedule and tight timing. To secure the smooth functioning, one more thing was needed: keeping up with the timetable. The congress organizers devised an original technical solution for ensuring this:

> In most of the lecture-rooms there were traffic-lights operated by the chairman in order to keep the speakers to the time allotted. Yellow light meant you can speak for another two minutes; red light meant stop.

The scientific program was complemented by three symposia, organized by the Wiskundig Genootschap in connection with the congress. The topics of the symposia were:

A. Stochastic Processes,

B. Algebraic Geometry,

C. Mathematical Interpretation of Formal Systems.

These symposia were run independently of the congress, although they were held almost simultaneously and their lectures were included in the proceedings of the congress (except for those of Symposium C).

The traditional exhibition of mathematical books was complemented with an exhibition of didactic and pedagogical works in mathematics, organized under the auspices of the International Commission on Mathematical Instruction.

There were also activities of a different character. Three of them stand out because of their novelty and singularity.

A demonstration was given of "electronic devices," namely:

- the IBM electronic calculators 604 and 626,
- the electronic computer ARRA built in the Mathematical Centre of Amsterdam,
- the electronic computer "Miracle" built by Ferranti and belonging to the Royal Shell Research Laboratories (B.P.M.) of Amsterdam.

Linen tablecloths with the Gaussian primes in the complex plane were sold at ICM 1954. (Courtesy of Sanny de Zoete, Holland.)

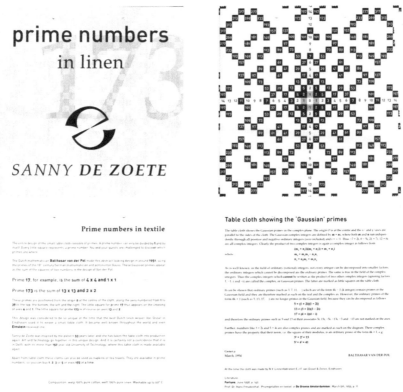

Explanation by Balthasar van der Pol of the Gaussian primes in the complex plane used for the linen tablecloths. (Courtesy of Sanny de Zoete, Holland.)

The best Dutch linen company, Van Dissel & Zn from Eindhoven, had woven dinner napkins in linen with a special design by Balthasar van der Pol: they showed the Gaussian primes in the complex plane. The tablecloths were displayed and sold during the congress at the Royal Tropical Institute, where the headquarters of the congress was located. (One can still buy them from the Sanny de Zoete, Antique & Design Linen from Delft.)

And last but not least, there was a superb exhibition at the Stedelijk Museum of Amsterdam of graphical work by the Dutch artist M. C. Escher. We will discuss this exhibition in detail in "Social Life at the ICM."

From the point of view of anniversaries, the year 1954 was an important one: it was the centenary of the birth of Henri Poincaré. Thus, independently, but closely connected with the congress, there was a commemorative session that took place in the Hague, presided over by Gaston Julia.

A short note on finances: among the donors for the congress, there were a variety of Dutch companies from varied sectors: Philips, De Bataasfsche Petroleum, Swets & Zeitlinger Booksellers, Unilever, Van Doorne's Automobielfabriek, and many others. The fee per participant was 50 Dutch guilder.

The closing session of the congress, also held at the Concertgebouw, witnessed another step in the increasing liaison between the International Mathematical Union and the congresses. This time, the topic was the procedure for deciding the venue of the next congress. A joint committee was chosen by the organizing committee and the union to make a recommendation. The proposal, recommended and received with great applause, was that presented by W. V. D. Hodge, inviting in the name of "the mathematicians from Great Britain and Northern Ireland," to hold the 1958 congress in Edinburgh.

There was also another proposal, by the Department of Mathematics of the Hebrew University in Jerusalem, but, since it concerned the 1962 congress, its consideration was forwarded to the next congress.

Proposal to hold the 1962 ICM in Jerusalem. (Courtesy of the Archives of the International Mathematical Union at the University of Helsinki.)

EDINBURGH 1958

ON THURSDAY, AUGUST 14, 1958, McEwan Hall hosted the inaugural session of the International Congress of Mathematicians (see also page 163). The congress was sponsored by the Royal Society of London, the Royal Society of Edinburgh, the University of Edinburgh, and the City of Edinburgh. It was placed under the patronage of H. R. H. Prince Philip, Duke of Edinburgh (who, not being able to attend the ceremony, sent a message from Buckingham Palace). The dates chosen for the congress were later than those of Amsterdam, due to the yearly celebration (since 1947) in early August of the now renowned Festival of Music and Drama of Edinburgh.

Opening of the 1958 congress at McEwan Hall of the University of Edinburgh. (Courtesy of the American Mathematical Society.)

W. V. D. Hodge, president of the 1958 ICM. (Courtesy of the American Mathematical Society.)

As for the previous congress, the chairman of the organizing committee, W. V. D. Hodge, was elected president of the congress at the proposal of the president of the Amsterdam congress (which was presented by an emissary). In his first presidential address, Hodge memorialized Sir Edmund Whittaker, who had passed away two years before.

The moment for the awarding of the Fields Medals had arrived. The president of the International Mathematical Union, Heinz Hopf from Zurich, chaired the medal committee consisting of Chandrasekharan from Bombay, Friedrichs from New York, Hall from Cambridge, Kolmogorov from Moscow, Schwartz from Paris, Siegel from Göttingen, and Zariski from Cambridge, MA.

The medals were awarded to

* Klaus Friedrich Roth from the University of London,
* René Thom from the Université de Strasbourg.

The medals were presented to the awardees by the Lord Provost of the City of Edinburgh. For the first time, reports on the work of the laureates were given by experts in each field. This occurred later in the congress. Davenport, replacing Siegel in this duty, reported on Roth's work in solving "the principal problem concerning approximation to algebraic numbers by rational numbers." Hopf explained Thom's creation of the theory of cobordism and stated that "for a long time, only few events have so strongly influenced Topology, and through Topology, other branches of mathematics as the advent of this work."

The inaugural session was adjourned, and the national anthem of the United Kingdom was played. In the afternoon, the Lord Provost, Magistrates, and Council of the City of Edinburgh sponsored a garden party for the members of the congress on the grounds of Lauriston Castle.

The scientific program followed the now well-established format: one-hour addresses by invitation, 19; half-hour addresses by invitation presented at the sections, 37; and fifteen-minute short communications, 604. Perhaps because of the large number of communications, and contrary to the practice of previous congresses, it was decided not to print their abstracts in the proceedings, since "it is better that they should follow the normal channels of publications." Instead, a booklet with the summaries of the communications was given at the congress.

The plenary addresses, in alphabetical order, were

* "Modern Development of Surface Theory," by A. D. Alexandrov;
* "On Some Mathematical Problems of Quantum Field Theory," by N. N. Bogolyubov and V. S. Vladimirov;
* *"Sur les fonctions de pluiseurs variables complexes: les espaces analytiques,"* by H. Cartan;
* *"La théorie des groupes algébraiques,"* by C. Chevalley;
* "Applications of Homological Algebra in Topology," by S. Eilenberg;
* "Some New Connections between Probability and Classical Analysis," by W. Feller;
* "Some Trends and Problems in Linear Partial Differential Equations," by L. Gårding;
* "The Cohomology Theory of Abstract Algebraic Varieties," by A. Grothendieck;
* *"Komplexe Mannigfaltigkeiten,"* by F. Hirzebruch;
* "Mathematical Logic: Constructive and Nonconstructive Operations," by S. C. Kleene;
* "Extended Boundary Value Problems," by C. Lanczos;
* "Optimal Processes of Regulation," by L. S. Pontryagin;
* "Rational Approximations of Algebraic Numbers," by K. F. Roth;
* "Extremum Problems and Variational Methods in Conformal Mapping," by M. Schiffer;
* "Cohomology Operations and Symmetric Products," by N. E. Steenrod;
* "Linearization and Delinearization," by G. Temple;
* *"Des variétés triangulées aux variétés différentiables,"* by R. Thom;

"Some Fundamental Problems in Statistical Physics," by G. E. Uhlenbeck;

"Entwicklungslinien in der Strukturtheorie der endlichen Gruppen," by H. Wielandt.

An innovation for this congress came in the list of sections: one section was devoted solely to topology, and a different emphasis was given to applied mathematics, highlighting the role of numerical mathematics. The sections were

- Section I: Logic and Foundations,
- Section II (a): Algebra,
- Section II (b): Theory of Numbers,
- Section III (a): Classical Analysis,
- Section III (b): Functional Analysis,
- Section IV: Topology,
- Section V (a): Algebraic Geometry,
- Section V (b): Differential Geometry,
- Section VI: Probability and Statistics,
- Section VII (a): Applied Mathematics,
- Section VII (b): Mathematical Physics,
- Section VII (c): Numerical Analysis,
- Section VIII: History and Education.

Inevitably, the modern world was reflected in the congress. In the short communications, we find traces of the changes that were occurring; for example, there was one talk presented in Section VIII entitled "Teaching Mathematics on Television."

It is interesting to highlight the exhibition of mathematical typography arranged by the Monotype Corporation at the Heriot-Watt College and the exhibitions of mathematical books in the Scottish National Library and the Library of the University of Edinburgh. It is a pity that the sparse proceedings of this congress do not give more information about these exhibits, which certainly were noteworthy.

In the large list of donors for the congress, there are many British companies, among them Babcock & Wilcox, British Aluminium, Imperial Chemical Industries, National Bank of Scotland, Rolls-Royce, Shell Petroleum, The Great Universal Stores Ltd., and many others. There was a participant's fee of five pounds.

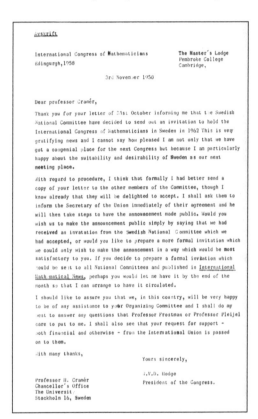

Swedish decision over the 1962 congress. (Courtesy of the Archives of the International Mathematical Union at the University of Helsinki.)

The closing session was held on the afternoon of Tuesday, August 21, at McEwan Hall. At the Amsterdam congress, a joint committee consisting of representatives of the International Mathematical Union and of the Edinburgh congress was formed to consider the location for the 1962 congress. Hodge made an unusual announcement to the congress:

> I am authorized by the committee to say that while for reasons of a technical nature it is not possible to make any announcement today of the name of the host country for 1962, the prospects of holding a congress in that year amount to a certainty.

For the closing of the congress, "instead of a banquet as in previous congresses," a congress reception was held in the Royal Scottish Museum.

STOCKHOLM 1962

Finally, after having presented the proposal for the first time in the Rome 1908 congress and having obtained the approval in 1912 in Cambridge for the 1916 congress, which was canceled because of the Great War, Mittag-Leffler's dream of holding an International Congress of Mathematicians in Stockholm came true in 1962.

Gösta Mittag-Leffler (1846–1927). Forty-two years later, Mittag-Leffler's dream came true: the ICM in Stockholm. (Courtesy of the Center for History of Science of the Royal Swedish Academy of Sciences.)

However, the path for this congress was not easy. The failure in announcing the venue of the 1962 congress at the closing session of the Edinburgh congress was the result of the existence of competing proposals from several countries. In order to avoid a difficult decision, the committee in charge decided to approach the Swedish mathematicians at the congress, who, caught by surprise and in view of the immense amount of work and large financial resources needed to undertake the task successfully, did not give a definite answer, although they showed some sympathy for the idea. Once the Swedish Mathematical Society and other bodies had given their endorsement, the congress was announced.

The task was still enormous, especially for the relatively small Swedish mathematical community. Thus, the organizing committee, chaired by Otto Frostman, decided to approach the International Mathematical Union in order to collaborate in developing the scientific program. In this manner, the Fields Committee was appointed, the one-hour invited speakers were chosen, and the chairmen of the international panels in charge of recommending the half-hour invited speakers were named.

The opening ceremony took place on Wednesday, August 15, at the Konserthuset, the concert hall of Stockholm, under the patronage and in the presence of King Gustav VI Adolf of Sweden. The inauguration was musical: an orchestra from the Royal Navy played a selection of Swedish music. At the proposal of

Opening ceremony of the 1962 congress in Stockholm. (Courtesy of the Center for History of Science of the Royal Swedish Academy of Sciences.)

The public at the opening ceremony of the 1962 congress in Stockholm. (Courtesy of the Center for History of Science of the Royal Swedish Academy of Sciences.)

King Gustav VI Adolf of Sweden presenting the 1962 Fields Medals. (Courtesy of the Center for History of Science of the Royal Swedish Academy of Sciences.)

Frostman, Rolf Nevanlinna, president of the International Mathematical Union, was elected to preside over the congress (the only precedent of sharing these two posts had been Salvatore Pincherle in the Bologna 1928 congress).

The Fields Medal Committee consisted of P. S. Alexandrov, Artin, Chern, Chevalley, Whitney, and Yosida, and was presided over by Nevanlinna. The medals were awarded to

- Lars Hörmander from the University of Stockholm,
- John Milnor from Princeton University.

The medals were presented to the awardees by the King. Lars Gårding spoke on the "outstanding work in partial differential equations" of Hörmander, and Whitney explained that "differential topology is a strong young field ... its vitality is largely due to the fine achievements of Milnor."

The scientific program of the congress followed the lines of that of Edinburgh. Indeed, the scientific sections had very few changes. The one change was combining differential geometry and topology and moving philosophy to Section I together with logic and foundations.

There were 16 one-hour invited addresses, 57 half-hour invited addresses delivered within the sections, and 745 short ten-minute communications. A "mimeographed booklet" containing outlines of the invited addresses was distributed at the congress, and, as in the Edinburgh congress, the abstracts of the short communications were confined to a booklet also given out at the congress. Most of the lectures were delivered at the Kungliga Tekniska Högskolan, the Royal Institute of Technology, allowing the participants to enjoy its beautiful campus.

The 16 plenary lectures were

- "Teichmüller Spaces," by L. V. Ahlfors;
- "Arithmetic Properties of Linear Algebraic Groups," by A. Borel;
- "Logic, Arithmetic, and Automata," by A. Church;
- "Markov Processes and Problems in Analysis," by E. B. Dynkin;
- "Homotopy and Cohomology Theory," by B. Eckmann;
- "Automorphic Funtions and the Theory of Representations," by I. M. Gelfand;

Attending lectures in the 1962 congress. (Courtesy of the Center for History of Science of the Royal Swedish Academy of Sciences.)

* *"Die Bedeutung des Levischen Problems für die analytische und algebraische Geometrie,"* by H. Grauert;

* "Problems of Stability and Error Propagation in the Numerical Integration of Ordinary Differential Equations," by P. Henrici;

* *"Transformées de Fourier des fonctions sommables,"* by J.-P. Kahane;

* "Topological Manifolds and Smooth Manifolds," by J. Milnor;

* "Geometrical Topology," by M. H. A. Newmann;

* "Some Aspects of Linear and Nonlinear Partial Differential Equations," by L. Nirenberg;

* "Algebraic Number Fields," by I. R. Shafarevich;

* "Discontinuous Groups and Harmonic Analysis," by A. Selberg;

* *"Géométrie algébrique,"* by J.-P. Serre;

* *"Groupes simples et géométries associées,"* by J. Tits.

Dynkin's and Gelfand's lectures were not presented by their authors, since they did not attend the congress; Kolmogorov and Mackey, respectively, read those lec-

tures. It is noteworthy that not one of the plenary speakers was a Swede.

Section VIII on education hosted several special meetings organized by the Mathematical Instruction Commission. Two of them are a good sample of the concerns of the time, when the reform of mathematics education was being undertaken:

* Which subjects in modern mathematics and which applications of modern mathematics can find a place in programs of secondary school instruction?

* Connections between arithmetic and algebra in the mathematical instruction of children up to the age of 15.

In the 1962 congress, Sergey L. Sobolev (right) gave a special report on the use of electronic computers at the University of Novosibirsk for deciphering the Maya language. (Courtesy of the Center for History of Science of the Royal Swedish Academy of Sciences.)

Computers and technical equipment found their spot in the congress in an exhibition arranged by IBM, Facit, and Ericsson, aimed at those participants interested in numerical methods and applied mathematics. But the most interesting complementary activity was the special report delivered by S. L. Sobolev on the "use of electronic computers for the deciphering of the Maya language made at the University of Novosibirsk."

The supporting bodies for the congress were the Swedish government, the International Mathematical Union, the City and University of Stockholm, the Mittag-Leffler Institute, and a variety of companies, among them Bolidens, Facit Electronics, Kockums Mekaniska, Scandinavian Airlines, Ericsson, and the publisher Almqvist & Wiksells. The fee for the congress was 160 Swedish kronor, "considerably higher than for the previous congresses, but in fact, quite normal in Sweden for a congress of this type and size," as was explained in the proceedings.

A new feature of the congress was that the organizers used the *World Directory of Mathematicians* for disseminating the announcements of the congress. The *Directory* was a long-standing project first outlined by Ferdinand Rudio in the Zurich 1897 congress. The project was relaunched after the first General Assembly of the International Mathematical Union in 1952. The task was a difficult one: gathering information (name, address, position, fields of interest) from all working mathematicians in the world. After a number of difficulties, the first edition was published in 1958 thanks to the effort of Komaravolu Chandrasekharan from Bombay, and the cooperation of the Tata Institute of Fundamental Research. New editions continued to be published every four years until publication was stopped in 2002 due to the advances in the use of electronic information.

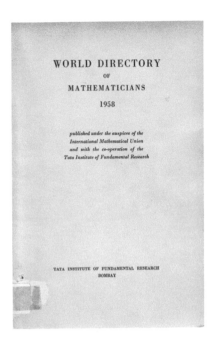

The first edition of the *World Directory of Mathematicians*. (Courtesy of the International Mathematical Union.)

The closing ceremony took place at the Konserthuset on Wednesday, August 22. President Nevanlinna was happy to acknowledge that the International Mathematical Union had become a definite partner of the international congresses. He announced that Georges de Rham had been elected president of the union for the period 1963–1966.

Academician Mikhail Alekseevich Lavrentyev invited the congress, in Russian, to convene in the Soviet Union in 1966 (an English translation was provided for the congress by Pavel Sergeyevich Alexandrov).

A PERIOD OF SUCCESS

WE END THE ACCOUNT of the international congresses during the 1950s and early 1960s by considering two issues. One is the participation of Soviet mathematicians, and hence of mathematicians from other socialist countries, in the international congresses; the other is the size of the ICM in this decade.

The organizing committee for the 1950 congress took care to avoid any political interference in the congress, and so it worked closely with the U.S. administration to secure visas for any mathematicians who wanted to attend the congress. Indeed, two important mathematicians at that congress had problems with their entry visa to the U.S.A.: Hadamard, honorary president, and Schwartz, Fields medalist. The Cold War had just begun; Schwartz had been a Trotskyist and Hadamard a sympathizer of the French Communist Party. Both cases were solved only by the direct intervention of the president of the United States, Harry Truman. In Hadamard's case, the issued was solved just a few days before the congress began.

After all these efforts, the organizers of the congress were proud to declare that only three people (one from an "independent country" (sic) and two from occupied countries) failed to attend because they could not get visas, and this happened because they had not adequately informed the congress organizers of their problems.

Soviet mathematicians had been absent from the international congresses since 1932. There had been no answer from the Soviet Union to any of the invitations to the 1950 congress or to the Constitutive Convention of the International Mathematical Union, and Kolmogorov, who had been appointed to the Fields Medal Committee, had not participated in the deliberations. The fact was that no mathematician from the Soviet Union or from any of the socialist countries (Bulgaria, Czechoslovakia, Hungary, Poland, and Rumania) who had attended the 1936 congress was present at the 1950 congress. There was one exception, Yugoslavia, which had already started to walk at its own pace.

However, just before the opening of the 1950 congress, a cablegram from Moscow was received; it was read at the opening:

USSR Academy of Sciences appreciates receiving kind invitation for Soviet scientist take part in International Congress of Mathematicians to be held in Cambridge. Soviet Mathematicians being very much occupied with their regular work unable to attend the congress. Hope that impending congress will be significant event in mathematical science. Wish success in congress activities.

S. Vavilov, President USSR Academy of Sciences

This friendly message opened the hope for future attendance of mathematicians from behind the Iron

Curtain at international congresses. Stalin's death in 1953 surely facilitated these hopes coming true.

Thus, at the 1954 congress, there were mathematicians from Bulgaria, 1; Czechoslovakia, 4; Hungary, 2; Poland, 6; Romania, 4; Soviet Union, 5; and Yugoslavia, 13. (There was even one mathematician who declared Latvian citizenship, but this is a different story.) The East German institutions also sent delegates, from the Karl-Marx Universität of Leipzig and the Deutsche Akademie der Wissenschaften zu Berlin. (In those times, the German mathematical representation was not yet split between the two German states.) However, despite the small number of participants present, these countries were very well represented among the plenary lecturers: four were from the Soviet Union (Alexandrov, Gelfand, Kolmogorov, and Nikolsky) and one from Poland (Borsuk).

In 1957, the Soviet Union and other socialist European countries joined the International Mathematical Union; so, for the Edinburgh 1958 congress, the Soviet delegation was the largest to date, 35 members.

Attendance from socialist countries to the Stockholm 1962 congress was close to normal (here we must take into account the high costs of the congress): USSR, 50; Poland, 50; Hungary, 40; Czechoslovakia, 25; Yugoslavia, 24; Romania, 15; and East Germany, 10. Of the 16 plenary speakers, three were from the Soviet Union. Due to the fairly large number of Russian speakers, a booklet containing translations of Russian texts was distributed at the congress.

Finally, Soviet attendance at the ICMs was normalized. The decision to hold the 1966 congress in the Soviet Union was definite proof of the Soviet interest in the international congresses.

The post World War II period was a time of major change in the world. Some of these changes can be detected through a close look at the international congresses.

Announcement in Russian of the Edinburgh 1958 congress. (Courtesy of the Archives of the International Mathematical Union at the University of Helsinki.)

The Cambridge, MA, 1950 congress had 1700 participants, the Amsterdam 1954 congress had 1553 (1436 men and 117 women), the Edinburgh 1958 congress had 1658, and there were 2107 at Stockholm in 1962. These impressive figures are more than double the highest figure from before World War II; that figure was attained at Bologna in 1928, when 836 mathematicians attended the congress. The organizers of the 1950 congress proudly declared: "The Congress was undoubtedly the largest gathering of persons ever assembled in the history of the world for the discussion of mathematical research."

But for the 1962 congress, if we also add up the associate members, that is, people accompanying the congress participants, 3091 persons attended. The organizers of the congress thus declared: "These figures, both separately and combined, exceed those for any of the previous congresses of mathematicians, as well as for any other scientific congress held in Sweden."

The 1950 congress was markedly North American: of the 1700 participants, 1410 were from the U.S.A. and Canada, and 290 from other countries. There were representatives from 173 universities and colleges of the U.S.A., and the only state not represented was South Dakota. Also, of the 22 plenary speakers, 15 were from the U.S. or came from U.S. institutions. Some hint of the then fashionable "American way of life" can be seen in the recommendation made in the announcements of the congress: "It is hoped that American mathematicians will be able to assist in the entertainment by putting their automobiles at the disposal of the entertainment committee for trips to be made out of Cambridge."

Unfortunately, the attendance of non-North Americans was less than expected by the organizers, who had made an extraordinary effort to provide transportation grants, room (in the dormitories of Harvard University), and board to all mathematicians from outside North America.

On the contrary, the Amsterdam 1954 congress was markedly European: of the 1553 participants, 1159 were European, 245 from North America, and 149 from other countries. The largest national groups were Great Britain, 261; the U.S.A., 228; the Netherlands, 212; and Germany, 207. It was similar for the Edinburgh 1958 congress: of the 1658 participants, only 360 were from the U.S.A.; the largest national group was the British with 500. The rest of the countries had smaller figures: France, 155; Germany, 150; Italy, 55 (unfortunately, the information from the 1958 congress is very scarce; there is not even a list of participants).

The figures for the Stockholm 1962 congress were more balanced: of the 2107 participants, the largest group was from the U.S.A. with 615, which together with the 57 Canadians constituted almost a third of the congress; for other European countries, attendance was United Kingdom, 302; West Germany, 155; Sweden, 116; France, 103; the Netherlands, 85; and Italy, 81. As we have seen, there was a fairly large group from the countries behind the Iron Curtain, 218.

At the 1950 congress, there were mathematicians from 40 countries (the proceedings declare 41, but because England and Scotland were listed as different countries). Two of the countries represented in the 1936 congress, Estonia and Latvia, were not listed in 1950 (they had been "absorbed" by the Soviet Union), and Palestine had turned into the newly created state of Israel. In Amsterdam, there were mathematicians from 51 countries (plus 12 participants who declare themselves to be "stateless," and one from Latvia). In Stockholm, there were 59 countries listed, but two of them were Armenia and Latvia, republics of the U.S.S.R. at the time, so technically there were 57. The independence of African countries is revealed with the presence of Cameroon, Nigeria, and Sierra Leone.

The very detailed data supplied by the organizers of the Amsterdam congress, distinguishing nationality and residency, reveals the effects of the war and the beginning of international mobility among mathematicians: of the 1553 participants, 1362 resided in the country of their nationality, and 91 did not. In the case of the United States, there were 242 participants from the U.S.A., of which only 198 had that nationality. We also witness the rise of new scientific institutions: in 1954, the Mathematischesforschungsinstitut from Oberwolfach/Baden sent a representative to the congress.

The dangers of a continuing increase in the size of the international congresses were discussed in the open-

Year	Participants	Countries	Lectures	Communications
1897	208	16	4	30
1900	250	26	4	33
1904	336	19	4	78
1908	535	22	10	127
1912	574	28	8	122
1920	200	27	5	79
1924	444	28	8	241
1928	836	36	14	419
1932	667	35	21	247
1936	487	36	19	205
1950	1700	40	22+20	374
1954	1553	51	20+42	496
1958	1658	–	19+37	604
1962	2107	57	16+57	745

Participation figures for the congresses up to 1962.

ing speeches of three of the presidents of the congresses: Veblen, Schouten, and Hodge.

The amount of work required for these large congresses was enormous, requiring the involvement of a large number of people. Consequently, the number of committees increased; there was a committee for the program, the finances, the proceedings, the sections, the budget, technical matters, accommodations, excursions, registration and reception, cooperation, entertainment, publicity, travel grants, etc.

Two positive consequences of the increasing size of the congress were that "students acted as stewards at headquarters, in the lecture rooms and elsewhere" and "mathematicians from all parts of the world acted as chairmen of the sessions." This shows that the holding of an ICM was a task in which progressively more groups of people related to mathematics were involved.

However, the debate on the difficulties caused by the size of the international congresses was somehow suspended by Hodge at the closing of the Edinburgh congress when he said:

Through the choice of the invited speakers and through the large number of communications of other members the congress has presented a picture of mathematics today and its trends. But the international congresses have another purpose, which I believe is just as important, that of promoting fellowship between mathematicians of all countries. This fellowship has its roots in our common love for our science, to whose growth we all try to contribute. It is the responsibility of each generation to take care that this fellowship is maintained and strengthened, and extended to the new generation.

We end this part devoted to the congresses in the 1950s and early 1960s with the panoramic group photograph of the Cambridge 1950 congress. Among the many interesting features of the photo (apart from its size: 25 by 75 centimeters), there is one that could almost be predicted in the announcements of the congress, that "special provisions are being made for the care of children." Note the number of children in the picture; we were already in the baby boom era.

Panoramic photograph of the 1950 congress in Cambridge, MA. (Courtesy of Harvard University Archives, Call # HUP-SF International Congress of Mathematicians.)

Interlude

BUILDINGS OF THE ICM

The relevance of the ICM as a social event in the countries and cities where the congresses have been held can be gauged by the importance of the buildings that have hosted the congresses. Universities, congress and convention centers, and other representative buildings reflect the history of changing architectural styles (Gothic, Classicist, Baroque, their historicist revivals in the nineteenth century, Rationalism, and the functional architecture of the twentieth century).

Zurich's Eidgenössische Polytechnikum. (Courtesy of Olli Lehto.)

Some of these buildings are based on projects of well-known architects, and they stand out in the history of architecture. Among these we find the Congress Palace of the Paris *Exposition Universelle*; the Tonhalle of Zurich; the Old University of Heidelberg; the University of Edinburgh, designed by Reginald Ely; the Lomonosov State University of Moscow, an impressive tower built under Stalin's rule; the Finlandia-talo, masterpiece by the influential Alvar Aalto; and the Palacio Municipal de Congresos of Madrid, by Ricardo Bofill.

The pictures included here present these and other congress buildings, in chronological order of the congresses.

Université de la Sorbonne. (© Olivier Jacquet–Université Paris-Sorbonne.)

Palais des Congrès, Exposition Universelle, Paris 1900. (From *Exposition Universelle de 1900,* Resengoti, 1900.)

Alte Universität Heidelberg. (Courtesy of the Universitätsarchiv Heidelberg.)

Palazzo Corsini, Rome.

King's College, Cambridge. (Courtesy of Wayne Boucher.)

Université de Strasbourg. (© Cordon Press.)

Convocation Hall, University of Toronto. (From the proceedings of the 1924 ICM, The University of Toronto Press 1928.)

101 - BOLOGNA - ARCHIGINNASIO (CORTILE)

Archiginnasio di Bologna. (Courtesy of the Biblioteca Comunale dell' Archiginnasio, Commune di Bologna.)

Eidgenössische Technische Hochschule, Zurich. (© ETH Zürich. Photo: Susi Lindig.)

The University of Oslo. (Courtesy of the University of Oslo, University History Photobase.)

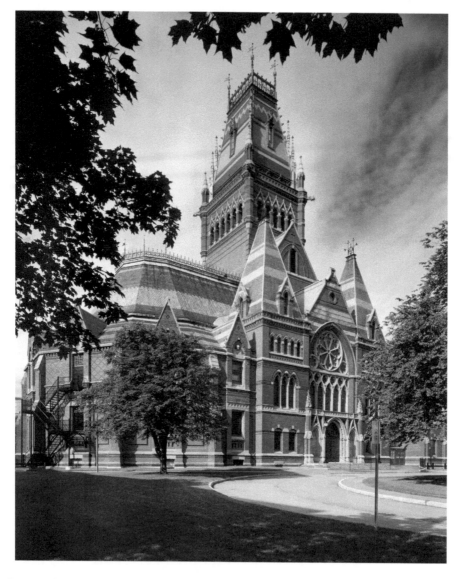

Memorial Hall of Harvard University. (Courtesy of the President and Fellows of Harvard College. Photo: Steve Rosenthal.)

Concertgebouw, Amsterdam. (© Radial Press.)

McEwan Hall, University of Edinburgh. (Courtesy of Edinburgh University Library, Special Collections Department, Phot. IU. 251.)

Stadshuset of Stockholm. (Courtesy of the Stadshuset Stockholm. Photo: Yanan Li.)

Lomonosov Moscow State University. (Courtesy of N. Molchanov.)

Frederick Wood Theatre, University of British Columbia. (Courtesy of the University of British Columbia Archives.)

The Finlandia Hall, Helsinki. (Courtesy of the Finlandia-talo. Photo: Rauno Träskelin.)

The University of Helsinki. (Courtesy of the University of Helsinki.)

The Palace of Culture of Warsaw. (Courtesy of the Instytut Matematyczny Polskiej Akademii Nauk.)

The campus of the University of California at Berkeley. (Courtesy of Steve McConnell / University of California, Berkeley.)

The Kyoto International Conference Hall. (Courtesy of the Kyoto International Conference Hall.)

Zurich's Kongresshaus. (Courtesy of Zurich's Kongresshaus.)

Eidgenössische Technische Hochschule, Zurich. (© ETH Zürich.)

The International Conference Center of Berlin. (Courtesy of the International Conference Center of Berlin.)

Mathematics building, Technische Universität Berlin. (Courtesy of the Technische Universität Berlin.)

The Great Hall of the People, Beijing. (Courtesy of Kikutake Yuji.)

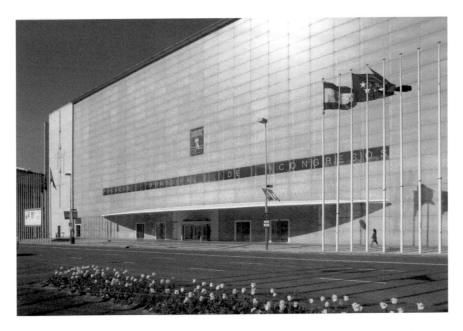

The Palacio Municipal de Congresos of Madrid. (© Carlos Casariego.)

PART IV
ON THE ROAD

IN A CERTAIN SENSE, the Moscow congress of 1966 gave new significance to the ICM that continues through the present time. Some of the reasons were its impact within the mathematical community, its size (it reached 4000 participants!), the involvement of a large number of Soviet mathematicians, its mathematical program, and also its resonance outside mathematics. It began a period in which the ICM again suffered from international tension, in this case as a result of the Cold War.

The congresses were held in

- Moscow, August 16–26, 1966;
- Nice, September 1–10, 1970;
- Vancouver, August 21–29, 1974;
- Helsinki, August 15–23, 1978.
- Warsaw, August 16–24, 1983;
- Berkeley, August 3–11, 1986.

Lomonosov Moscow State University, venue of the Moscow 1966 congress. (Courtesy of N. Molchanov.)

MOSCOW 1966

I have the great pleasure of greeting all congress participants in the name of the Academy of Sciences of the Soviet Union and transmitting the wish that the work of the congress will be fruitful.

ACADEMICIAN Mstislav Vsevolodovich Keldysh, president of the Academy of Sciences of the U.S.S.R. (and chief theoretician of Soviet cosmonautics in the 1960s), gave this socialist greeting when he opened the International Congress of Mathematicians in Moscow on August 16, 1966, at 4 p.m., in the modern Palace of Congresses, which had been recently built inside the Kremlin. The Rector of the Moskovskij Gosudarstvennyj Universitet—the (Lomonosov) Moscow State University (see also page 173), in short MGU—academician Ivan Georgievich Petrovsky, presided over the congress and also chaired the organizing committee (consisting also of academician Ivan Matveevich Vinogradov, academician Mikhail Alekseevich Lavrentyev, Sergey Mergelyan, and other Soviet mathematicians).

The congress provided a good opportunity for the Soviet system to show, in front of a worldwide scientific community, its achievements after the de-Stalinization process. The power of the country was at its zenith: just recently Yuri Gagarin had been the first man to travel into outer space and Valentina Tereshkova the first woman (in the race for conquering the cosmos,

Soviet cosmonauts seemed to be winning). The public image was carefully considered for the Soviet people: the congress held a daily press conference.

Soviet symbols in the Lomonosov Moscow State University, venue of the 1966 congress. (Courtesy of Kudelkin Nikolay, Moscow.)

There was a record-breaking attendance at the Moscow congress. According to Soviet statistics (not included in the proceedings), there were around 4280 mathematicians attending (there had been almost 5600 preregistered!). The impact of the congress can be gauged by looking at the number of Soviet mathematicians who attended: 1470. For a scientific community that had been very much isolated since the 1930s, being able to meet, discuss, and chat with over 2000 mathematicians from the West was an important occasion. (Attendance from the other side of the Iron Curtain was from the U.S.A., 725; from Great Britain, 286; from France, 280; from West Germany, 169 (the official list records separately 147 from the Federal Republic of Germany and 22 from West Berlin); from the Netherlands, 92; from Sweden, 89; from Canada, 83; and from Italy, 70.) The delegations from "brother socialist" countries were also large: 229 from East Germany, 120 from Poland, 94 from Hungary, 88 from Romania, 81 from Bulgaria, and 60 from Czechoslovakia. The overall number of countries represented was 54. Remarkably, attendance at the congress had doubled that of the Stockholm congress, which was the highest to date, and it was 20 times that of the first congress in Zurich in 1897.

Among the thousands of participants, there was an unexpected one, holding registration number 4397:

Nicolas Bourbaki. The question continuously circulating around the congress was: "Has Bourbaki already arrived in Moscow?"

Австралия	20 делегатов	Марокко	2 делегата
Австрия	3 »	Мексика	4 »
Алжир	3 »	Нигерия	3 »
Аргентина	1 »	Нидерланды	92 »
Афганистан	2 »	Новая Зеландия	5 »
Бельгия	9 »	Норвегия	17 »
Болгария	81 »	ОАР	1 »
Бразилия	3 »	Польша	120 »
Великобритания	286 »	Румыния	88 »
Венгрия	94 »	Сенегал	3 »
Вьетнам	5 »	Сирия	1 »
Гана	3 »	СССР	1479 »
ГДР	229 »	США	725 »
Греция	14 »	Танзания	1 »
Дания	17 »	Тунис	1 »
Замбия	1 »	Турция	3 »
Западный Берлин	22 »	Финляндия	15 »
Израиль	20 »	Франция	280 »
Индия	12 »	ФРГ	147 »
Иран	3 »	Чехословакия	60 »
Ирландия	9 »	Чили	5 »
Испания	7 »	Швейцария	58 »
Италия	70 »	Швеция	89 »
Канада	83 »	Югославия	46 »
КНДР	2 »	Япония	26 »
Куба	3 »		
Ливан	2 »		
Мадагаскар	2		

Membership by country in the Moscow 1966 congress. (From *Vsemirnii Kongress Matematikov v Moskve* by V. N. Trostnikov, Znanie 1967.)

Georges de Rham, president of the IMU, at the opening of ceremony of the Moscow 1966 congress. At his side are Henri Cartan and Ivan G. Petrovsky. (Photograph taken at the 1966 ICM in Moscow by S. V. Smirnov, from Ivanovo State University.)

General atmosphere at the opening of the Moscow 1966 congress. (Photographs taken at the 1966 ICM in Moscow by S. V. Smirnov, from Ivanovo State University.)

We have already explained that it was at this congress where, for the first time, four Fields Medals were awarded and the "under forty" rule was made explicit. Georges de Rham, as president of the International Mathematical Union, was chairman of the Fields Medal Committee, which was made up of H. Davenport, M. Deuring, W. Feller, M. A. Lavrentyev, J.-P. Serre, D. C. Spenser, and R. Thom. He announced the 1966 laureates, who were

- Michael Francis Atiyah from Oxford University,
- Paul J. Cohen from Stanford University,
- Alexander Grothendieck from the Université de Paris,
- Stephen Smale from the University of California at Berkeley.

The medals were presented to the awardees by academician Keldysh. Two absences somewhat dimmed the solemn moment: Smale arrived at the ceremony late, just in time to listen to part of the laudation of his work, and Grothendieck had announced that he would not attend the congress. In any case, Smale received his medal, and Grothendieck's was accepted on his behalf by Léon Motchane, founder and director of the Institut des Hautes Études Scientifiques, the IHES.

Afterwards, Henri Cartan spoke on Atiyah's achievements on *"la K-théorie, la formule de l'indice, et la formule de Lefschetz"*; Alonzo Church spoke on Cohen's solution to the continuum problem (which was Hilbert's first problem); Jean Dieudonné on Grothendieck's role in the renovation of algebraic geometry, *"débarrassées des restrictions parasites"*; and René Thom on Smale's *"grand travaux de 1960 sur la conjecture de Poincaré."*

For this congress, the scientific program was fully developed by the International Mathematical Union, although in close connection (it seems that more often it was in close disagreement) with the Soviet organizing committee. A similar structure to that of other post-war congresses was adopted: plenary addresses, invited addresses in the sections, and communications; see http://ershov.iis.nsk.su/archive/eaimage.asp?lang=1&did=30581&fileid=176629.

There were 17 plenary addresses, of which five were by U.S. mathematicians and five by Soviet mathematicians. They were

- "A Survey of Homotopy-Theory," by John F. Adams;

- "The Etale Topology of Schemes," by Michael Artin;

- "Global Aspects of the Theory of Elliptic Differential Operators," by Michael F. Atiyah;

- "Dynamic Programming and Modern Control Theory," by Richard Bellman;

- "Convergence and Summability of Fourier Series," by Lennart Carleson;

- "Hyperbolic Problems in the Theory of Surfaces," by N. V. Efimov;

- "Harmonic Analysis on Semisimple Lie Groups," by Harish-Chandra;

- "Analytic Problems and Results in the Theory of Linear Operators in Hilbert Space," by M. G. Krein;

- "Théorie locale des fonctions différentiables," by B. Malgrange;

- "On Some Questions on the Border of Algebra and Logic," by A. I. Malcev;

- "Automorphic Functions and Arithmetic Groups," by I. I. Piatetski-Shapiro;

- "Ungleichungen und Fehlerabschätzungen," by Johann Schröder;

- "Neuere Ergebnisse der Beweistheorie," by Kurt Schütte;

- "Differentiable Dynamical Systems," by Stephen Smale;

- "Some Recent Developments in Mathematical Statistics," by Charles M. Stein;

- "Characterization of Finite Simple Groups," by John G. Thompson;

- "Recent Developments in Analytic Number Theory," by I. M. Vinogradov and A. G. Postnikov.

One of the highlights of the program was Carleson's lecture where he presented the results of his paper published in *Acta Mathematica* in 1966 with the solution to the longstanding problem posed by the Russian mathematician Nikolai Lusin in 1913 regarding the pointwise convergence of Fourier series.

CONVERGENCE AND SUMMABILITY OF FOURIER SERIES

LENNART CARLESON

Let me first state quite explicitly that I do not intend to give in this lecture any survey of the very large field covered by the title. There is also no need for this since the Congress was presented such a survey quite recently. I rather want to present my personal interests which are concentrated on the almost everywhere behaviour of the partial sums. Also the subject of summability will only be touched upon.

1. Background

For a very long time, the outstanding result in the area of almost everywhere convergence has been the following result of Kolmogorov-Seliverstov-Plessner: *if for* $\lambda_n = \log n$

$$(1.1) \qquad \sum_1^\infty (a_n^2 + b_n^2)\lambda_n < \infty,$$

then

$$(1.2) \qquad s_n(x) = \frac{a_0}{2} + \sum_1^n (a_\nu \cos \nu x + b_\nu \sin \nu x)$$

converges a.e. The outstanding question was whether $\log n$ is a relevant sequence or not.

It has been known that conditions of the type (1.1) are related to capacities with respect to a kernel

$$(1.3) \qquad K(x) \sim \sum \frac{\cos nx}{\lambda_n}$$

(Beurling [1]: $\lambda_n = n$, $K(x) \sim \log\frac{1}{|x|}$; Salem-Zygmund [4]: $\lambda_n = n^\alpha$, $K(x) \sim |x|^{\alpha-1}$). However, what they really prove is that the capacity of the divergence set vanishes for

$$K^*(x) = \frac{1}{|x|}\int_0^{|x|} K(t)\, dt$$

(see Temko [5]). When $\lambda_n = (\log n)^\beta$,

$$K(x) \sim |x|^{-1}\left(\log\frac{1}{|x|}\right)^{-1-\beta}$$

6*

Lecture by Lennart Carleson in the 1966 congress. (From the proceedings of the 1966 ICM, Mir 1968.)

There were 64 invited addresses related to the sections (only 54 were published). The list of sections was refined, and its number increased to 15, almost doubling those of the previous congress. New topics appeared in the list, such as differential equations, ordinary or partial, and control theory. The new list of sections is given below (the number of published invited addresses within that section are given in parentheses, four of them not counted due to the late arrival of manuscripts):

- Section 1: Mathematical Logic and Foundations of Mathematics (2),

- Section 2: Algebra (4),

- Section 3: Theory of Numbers (2),

- Section 4: Classical Analysis (5),

- Section 5: Functional Analysis (3),

- Section 6: Ordinary Differential Equations (3),

- Section 7: Partial Differential Equations (4),

- Section 8: Topology (4),

- Section 9: Geometry (1),

- Section 10: Algebraic Geometry and Complex Manifolds (7),

- Section 11: Probability Theory and Statistics (2),

- Section 12: Applied Mathematics and Mathematical Physics (4),

- Section 13: Mathematical Problems of Control Theory (3),

- Section 14: Numerical Mathematics (4),

- Section 15: History and Pedagogical Questions (1).

There were also communications, but the extremely limited proceedings of the congress do not give any detail about the overall number, distribution of topics, authors, or titles; we only know that the organizers received 2100 proposals for communications. There is also no list of participants. This is a pity because we get a blurred image of the congress.

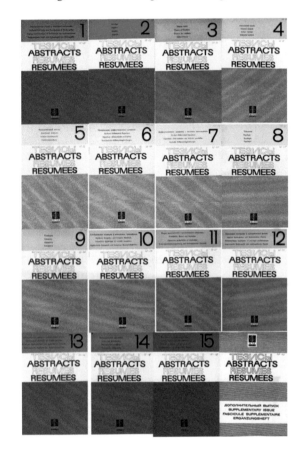

The scientific sections of the 1966 congress. (Courtesy of Academician A. Ershov archive, http://ershov.iis.nsk.su/archive/eaindex.asp?lang=2&did=30589.)

The lectures were held at the MGU situated on Lenin Hill. It is an impressive skyscraper of almost 40 stories, 240 meters high, one of seven similar buildings constructed in Moscow in the 1950s, popularly known

Busts of Lobachevsky and Chebyshev on the campus of Moscow University. (Courtesy of Kudelkin Nikolay, Moscow.)

by Muscovites as the Seven Sisters (see pages 164 and page 173). On the campus of the MGU, we can still find the busts of the renowned Russian mathematicians Nikolai Lobachevsky and Pafnuty Chebyschev. Participants recall constantly going up and down in the elevators in order to attend lectures in different sessions. The number of simultaneous communications delivered rose to 40. Soviet reports say that the total number of hours devoted to mathematics corresponded to 25 continuous days and nights!

Механико-математический факультет МГУ

приглашает Вас

НА ВЕЧЕР-ВСТРЕЧУ

с молодыми математиками

Вечер состоится 22 августа
в Актовом зале МГУ
(Ленинские горы)

Начало в 19 часов

Department of Mechanics and Mathematics
of Moscow University

invites you

TO A MEETING PARTY

with young mathematicians

The party will be held in the Assembly
Hall of Moscow University (on Lenin Hills)

on August 22 at 7 p. m.

Invitation for a meeting at the 1966 congress. (Courtesy of Academician A. Ershov archive, http://ershov.iis.nsk.su/archive/eaimage.asp?lang=2&did=30576&fileid=176667.)

Lectures at the Moscow 1966 congress. On the right, Andrei N. Tikhonov. (Photographs taken at the 1966 ICM in Moscow by S. V. Smirnov, from Ivanovo State University.)

Among the many events organized by the congress was a special lecture by academician Andrey Nikolaevich Kolmogorov on "Mathematical Education in Secondary Schools." Although the lecture was not part of the congress, attendance was so high that the lecture room had to be changed to a larger one. Another, this time peculiar, event was the "Meeting Party with Young Mathematicians" held in the Assembly Hall of Moscow University.

Unexpectedly, the congress made newspaper headlines, not for matters of a scientific nature (that would occur 40 years later, at the Madrid 2006 congress) but for political reasons.

The first event of a political nature was, or seems to have been, Grothendieck's refusal to attend the congress at which he was going to be awarded the Fields Medal. Although there are too many different versions of the reasons behind his decision (and no public statement from Grothendieck himself), apparently, one way or another, it was connected to his criticism of Soviet politics, either in relation to the treatment given to Soviet dissident writers or to the military interventions in Eastern Europe. In any case, these reasons for Grothendieck's refusal were not a matter of any news-

paper headline. (What was highlighted by the Soviet media was that Grothendieck had declared himself a citizen of the world and requested United Nations citizenship.)

What did reach the newspapers was the so-called Smale incident. The story, as Smale has explained it himself, goes as follows. While Smale was traveling to Moscow, the House Committee on Un-American Activities, a committee of the House of Representatives of the U.S.A., presented him with an official summons to testify in connection with his activities against the Vietnam War. When this became known at the international congress, a petition immediately circulated against the war and backing U.S. academics who opposed it. Smale was asked by a North Vietnamese reporter for an interview. Somehow, Smale was able to turn the interview into a press conference, so that U.S. reporters could also attend and give a faithful report of his words.

The press conference took place on the steps of Moscow University on the morning of Friday, August 26. There, Smale read a statement in which he said:

I believe the American Military Intervention in Vietnam is horrible and becomes more horrible every day.

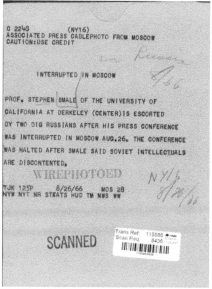

Steven Smale, a 1966 Fields medalist, gave a press conference on the steps of Moscow University, criticizing United States bombings in Vietnam and Soviet intervention in Hungary. (© Associated Press.)

I have great sympathy for the victims of this intervention, the Vietnamese people. However, in Moscow today, one cannot help but remember that it was only 10 years ago that Russian troops were brutally intervening in Hungary and that many courageous Hungarians died fighting for their independence.

The next scene was like something from a John Le Carré novel: Smale, after speaking with Soviet congress officials, was led into a car accompanied by two Soviet individuals; the car drove off at high speed and disappeared. Some time later, he was returned to the congress, where the closing ceremony was already taking place. What had happened in that unscheduled and unrequested car ride? Smale had been invited to see the Moscow museums, although in fact he was taken to the headquarters of the Soviet news agency. He remembers: "I felt pressured and a little scared. But all the while I was treated not just politely, but like a dignitary." The incident appeared in the *New York Times*, *Washington Post*, and other newspapers. The Soviet media version stressed Smale's political activism, the attempts of the FBI to prevent him from attending the Moscow congress, but did not mention his criticisms in the press conference.

Fortunately, neither of these incidents tainted the general atmosphere of the congress, and president de Rham, at the closing ceremony held in the Assembly Hall of the MGU, was able to praise the "cooperation between mathematicians of the Soviet Union and those of other countries, especially of Western Europe and the U.S.A." He announced that Henri Cartan had been elected president of the union for the period 1967–1970. This occurred at the assembly of the International Mathematical Union held just prior to the congress in the city of Dubna (a famous town near Moscow, hosting many important scientific institutions). A first Soviet proposal for the location for that meeting had been a sanatorium of the Academy of Sciences, but the offer was rejected by the union.

U.S. delegation at the Moscow 1966 congress. (Photograph taken at the 1966 ICM in Moscow by S. V. Smirnov, from Ivanovo State University.)

The venue for the 1970 congress was announced by Jean Dieudonné in his capacity as Dean of the Faculty of Sciences of the Université de Nice:

> The city of Nice, due to its location, its activity, its tourist facilities, and the existence of an active university, presents the required conditions to hold a scientific congress. I then propose that the International Congress of Mathematicians convenes in 1970 in the city of Nice.

The large-scale contact between mathematical communities that had been previously separated was one of the most valuable achievements of the congress. The atmosphere was casual and friendly. One could see groups of people sitting on the steps or playing badminton in the halls. The last day there was a soccer game, the U.S.S.R. team against the rest of the world; the U.S.S.R. won 5 to 2. Many Soviet mathematicians, young at the time, still have vivid memories of the congress: the excitement of the MGU full of foreign mathematicians. Moreover, it was also the longest congress to date: it lasted eleven days! In this sense, the Moscow congress was a landmark for the international congresses.

FEES

Mathematician : Fr : 200.— Guest : Fr : 100.—

Payment of the fee of Fr : 200.— gives you the right to receive all the Congress publications :
1°) — Individual written communications retained by the Organization Committee, and compiled in one volume which is given to each participant on his arrival at Nice.

2°) — Texts of lectures given by invitation of the Organization Committee which will be compiled into several volumes published after the Congress (in theory in 1971) whose total number of pages is estimated at about 2000.

PLEASE DO NOT SEND ANY FEES TO THE CONGRESS SECRETARIAT.
FURTHER ANNOUNCEMENTS WILL ONLY BE SENT TO PERSONS WHO HAVE PAID THE FEE.

All persons who have not paid the fee by 1st April 1970 will be required to pay an additional 10 %.

ROOM RESERVATIONS

See the subjoined registration form.

RENTING OF ACCOMMODATION

If you would like some information about renting an appartment or a villa for a period exceeding the duration of the Congress, you are asked to mention this on the Registration form beneath the heading " Special requests ". This information is furnished without guarantees and does not entail any responsability on the part of the Congress Organization Committee.

EXCURSIONS

See Registration form.

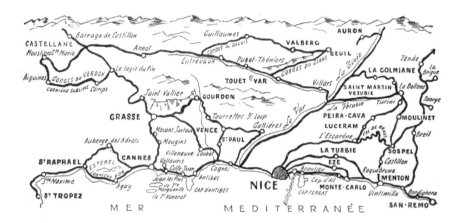

Announcement of the Nice 1970 congress. (Courtesy of the Archives of the International Mathematical Union at the University of Helsinki.)

NICE 1970

THE CÔTE D'AZUR, on the southeastern Mediterranean coast of France, with the famous cities of St. Tropez, Cannes, Nice, Monte Carlo, and San Remo, enjoys a worldwide reputation for first-class summer vacations. In order to avoid the height of the tourist season, the 1970 congress took place at the beginning of September. The Palais des expositions de la Ville de Nice was the venue for the inaugural session held on

Paul Montel (1876-1975), honorary president of the 1970 congress. (From *Selecta: 1897–1947*, P. Montel, Gauthier-Villars 1947.)

Tuesday, September 1. Henri Cartan, the president of the International Mathematical Union, proposed electing Jean Leray to preside over the congress (he had chaired the organizing committee), and Paul Montel, who was 94 years old at that time, was named honorary president.

Cartan had presided over the Fields Medal Committee, which included J. L. Doob, F. Hirzebruch, L. Hörmander, S. Iyanaga, J. Milnor, I. R. Shafarevich, and P. Turán. The four awardees chosen by the committee were

- Alan Baker from Cambridge University,

- Heisuke Hironaka from Harvard University,

- Sergei Novikov from Moscow State University,

- John G. Thompson from the University of Chicago.

The medals were presented by the French Minister of National Education. Unfortunately, Novikov was not able to be present at the ceremony to receive his medal.

Turán reported on Baker's work on the theory of transcendental numbers and the applications to Diophantine equations and commented that "besides the worthy tendency to start a new theory in order to solve a problem it pays also to attack specific difficult problems directly." Grothendieck talked about Hironaka's

CONGRÈS INTERNATIONAL
DES MATHÉMATICIENS

THE INTERNATIONAL CONGRESS
OF MATHEMATICIANS

INTERNATIONALEN
MATHEMATIKERKONGRESSES

Nice 1ᵉ 10 Septembre 1970

Fields 賞 授賞式, 9月18, 1970年
広中平祐
日本

ALAN BAKER
CAMBRIDGE 大学, ENGLAND

JOHN G. THOMPSON
CHICAGO 大学, USA

The 1970 Fields medalists (Sergei Novikov did not attend the congress).

result on the resolution of singularities of algebraic varieties and noted that "the result is not fully platonic ... on the contrary it is a powerful tool." Atiyah praised Novikov's "great originality and very powerful technique both in its geometric and algebraic aspects" in his work on geometric and algebraic topology. Thompson's contribution to the classification of finite simple groups was discussed by R. Brauer.

This congress exhibited very innovative features with regard to the scientific program. In the morning of each congress day, the only activity was two consecutive, one-hour, general lectures by invited speakers addressed to all participants. There were 16 of these general lectures:

* "Effective Methods in the Theory of Numbers," by A. Baker;

* "On the Topological Obstructions to Integrability," by R. Bott;

* "Manifolds and Homotopy Theory," by W. Browder;

* "Differential Geometry: Its Past and Its Future," by S.-S. Chern;

* "The Current Situation in the Theory of Finite Simple Groups," by W. Feit;

* "The Cohomology of Infinite Dimensional Lie Algebras: Some Questions of Integral Geometry," by I. M. Gelfand;

UNIVERSITÉ DE NICE
FACULTÉ DES SCIENCES
PARC VALROSE
N I C E   Nice, le July 8 , 1970

Dear Professor Frostman ,

                I have received verbal assuran-
ces that you will be able to get through French
customs without any trouble ; enclosed is a copy
of a letter which I sent to the director of the
customs Office in Nice . In any case , I will be
at the airport to make sure that everything is
OK .

          Sincerely yours

                       J.Dieudonné

ICM's delicate issues: Fields Medals going through customs. (Courtesy of the Archives of the International Mathematical Union at the University of Helsinki.)

- "A Transcendental Method in Algebraic Geometry," by P. A. Griffiths;
- "Linear Differential Operators," by L. Hörmander;
- "Scattering Theory and Perturbation of Continuous Spectra," by T. Kato;
- "Model Theory," by H. J. Keisler;
- "Methods and Problems of Computational Mathematics," by G. I. Marchuk;
- *"Les jeux différentiels linéaires,"* by L. Pontryagin;
- "Some Problems in Harmonic Analysis Suggested by Symmetric Spaces and Semi-simple Groups," by E. M. Stein;
- "Algebraic K-theory," by R. G. Swan;
- "Symbols in Arithmetic," by J. Tate;
- "Geometric Topology: Manifolds and Structures," by C. T. C. Wall.

Afternoons were devoted to specialized expository talks of 50 minutes, delivered by experts in each field. There were up to ten parallel sessions running for three hours (so that participants could chose to listen to three of them). These expository talks were invited talks, and there were 242 of them, a very large number.

The program was complemented by the possibility offered to groups of congress participants to procure a lecture room for mathematical meetings not scheduled in the official program.

However, all this time for invited lectures and room usage had to come from somewhere: it was decided that there would be no verbal short communications. The communications presented—only 265, a very small number when compared with previous congresses—were printed in a booklet and distributed.

The expository lectures were organized in 34 sections, which were grouped into six major topics:

A. Logique Mathématique (7),

B. Algèbre (46),

C. Géometrie et Topologie (52),

D. Analyse (83),

E. Mathématiques Appliquées (49),

F. Histoire et Enseignements (5).

The numbers in parentheses indicate the number of invited expository talks. Each section was subdivided into subsections, except for Section A on Logique Mathématique. The Algèbre section had the following six subsections:

B_1. Algèbre générale (7),

B_2. Catégories; algèbre homologique (7),

B_3. Groupes Finis (8),

B_4. Corps locaux et globaux; analyse p-adique (6),

B$_5$. Géométrie algébrique (10),

B$_6$. Théorie des nombres, élémentaire et analytique (8).

The Géometrie et Topologie section was subdivided into the following five subsections:

C$_1$. Topologie générale et algébrique (9),

C$_2$. Topologie des variétés (14),

C$_3$. Geométrie différentielle (5),

C$_4$. Analyse sur les variétés (11),

C$_5$. Groupes algébriques, fonctions automorphes et groupes semi-simples (13).

The Analyse section was subdivided into the following 12 subsections:

D$_1$. Espaces vectorielles topologiques (3),

D$_2$. Algèbres d'opérateurs; représentations des groupes localment compacts (8),

D$_3$. Théorie spectrale (4),

D$_4$. Algèbres de fonctions; analyse de Fourier (6),

D$_5$. Théorie du potentiel; processus de Markov (8),

D$_6$. Probabilités, théorie de la mesure, intégration (6),

D$_7$. Fonctions analytiques d'une variable complexe (2),

D$_8$. Fonctions et espaces analytiques complexes (9),

D$_9$. Ensembles exceptionnels en analyse (3),

D$_{10}$. Analyse fonctionnelle et équations aux dérivées partielles linéaires (16),

D$_{11}$. Analyse fonctionnelle et équations aux dérivées partielles non linéaires (9),

D$_{12}$. Systèmes dynamiques et équations différentialles ordinaires (8).

The Mathématiques Appliquées section was subdivided into the following eight subsections:

E$_1$. Aspects mathématiques de la théorie quantique des champs (4),

E$_2$. Théorie de la relativité (5),

E$_3$. Problèmes mathématiques de la mécanique du continu (13),

E$_4$. Théorie de contrôle optimal (7),

E$_5$. Combinatorie et algèbre finie (7),

E$_6$. Statistique mathématique (4),

E$_7$. Problèmes mathématiques de la théorie de l'information, langage machine (5),

E$_8$. Analyse numérique (4).

And the section on Histoire et Enseignements was subdivided into its two natural subsections.

Surprisingly, this scientific program was radically different from those of other congresses. Instead of a linear list of sections, there was a tree-like structure. How could this have been so devised? As explained in the proceedings of the congress, an international committee had been constituted, and 33 specialized commissions had been appointed to make recommendations for the invited lectures. (Except for the president, Adrian Albert, the members of the committee and of the commissions and their recommendations were confidential.) In this regard, the French organization had, to some extent, avoided the rules that the International Mathematical Union had been trying to implement.

However, the above facts still do not explain the singularity of the scheme of the scientific program and the peculiar structure of the sections. What is the key?

Well, here is a conjecture. Looking in detail, we appreciate that the scheme of sections clearly aims at presenting mathematical activity in a systematic way, proposing a division of mathematics into main areas, each of them subdivided into subareas. Indeed, the Nice scheme proposes a certain "architecture of mathematics." This was in stark contrast to the unstructured and dispersed nature of the enumerative lists of sections of the previous congresses, in particular that of

Moscow. At this point, it should be recalled that one of the programmatic texts of the Bourbaki group was *L'architecture des mathématiques*, published in 1948. Now, if we look at the members of the organizing committee of the congress, we find, among others, François Bruhat, Jean Dieudonné, Laurent Schwartz, Jean-Pierre Serre, and André Lichnerowicz (who was also the "father" of the French reform on the teaching of mathematics in the 1960s and 1970s). All were members of the Bourbaki group; Dieudonné was even one of the founding members, as was Henri Cartan, president of the International Mathematical Union during the preparation of the congress. Grothendieck, who was in charge of the laudation of Hironaka's work, was also a member of the Bourbaki group.

These facts lead to the possible interpretation of the Nice 1970 congress as the "Bourbaki ICM" and its section structure as the result of applying the Bourbaki program to the international congresses.

Among the expository lectures, there was one entitled "Diophantine Representation of Recursively Enumerable Predicates," presented in Section A on Logique Mathématique by Yuri V. Matiyasevich, a 23-year-old mathematician from Leningrad. In that lecture, Hilbert's tenth problem (stated below) was solved, in the negative:

> Given a Diophantine equation with any number of unknown quantities and with rational integral numerical coefficients: To devise a process according to which it can be determined by a finite number of operations whether the equation is solvable in rational integers.

The section on Enseignement had little activity. One of the reasons was that the International Commission on Mathematical Instruction had held in August 1969 in Lyon the first International Congress on Mathematical Education, in short ICME. Since the president of the ICME, Hans Freudenthal, had not informed the IMU (which considered the ICME as a subcommission of the union) of these plans, there was some resentment. However, the ICME became well accepted and has been held ever since every four years: 1972 in Exeter, 1976 in Karlsruhe, 1980 in Berkeley, 1984 in Adelaide, 1988 in Budapest, 1992 in Québec, 1996 in Seville, 2000 in Tokyo/Makuhari, 2004 in Copenhagen, 2008 in Monterrey, Mexico.

The Paris 1900 congress had witnessed a discussion of the idea of a universal scientific language. The passing of years had decided the issue. Despite the French organization's effort to use that country's language, English was clearly the predominant language at the congress: of the 16 plenary speakers, all spoke in English except for one (Pontryagin, from the U.S.S.R.); and of the 242 specialized, invited expository talks, 190 were delivered in English, 49 in French, and 3 in German. That is, almost 80 percent of the lectures were

Lecture by Yu. V. Matiyasevich on the solution to Hilbert's tenth problem. (From the proceedings of the 1970 ICM, Gauthier-Villars 1971.)

delivered in English. Let us just recall that in the Paris 1900 congress, all plenary speakers spoke in French, and in Heidelberg in 1904, each plenary speaker spoke his native language: one in English, one in French, one in German, and one in Italian. In the Zurich 1932 congress, of the 268 plenary and invited lectures, 112 were delivered in French, 41 percent; 95 in German, 35 percent; 27 in English, 14 percent; and 24 in Italian, 9 percent. What could have been considered as a problem before was now just a fact: the driving forces of the world had changed.

The congress was under the high patronage of the President of the French Republic, Georges Pompidou, and under the patronage of the Prime Minister, Jacques Chaban-Delmas. This was clearly useful for receiving the 378,000 francs from the French administration; another 15,000 francs came from the university, and 162,000 from a "committee to support the dissemination of the works of the congress." On this committee there were companies like Air France, L'Air Liquide, Alcatel, Banque de l'Indochine, Ciments Lafarge, Saint-Gobain-Pont-à-Mousson, Dunlop, Esso-Chimie, Gervais-Danone, Kodak-Pathé, L'Oreal, Rhône-Poulenc, Solvay, Thomson, and Springer-Verlag. But this was just barely half of the budget. The other half came from the fees of the 2811 participants, coming from 60 countries, which provided 576,000 francs.

Dieudonné took advantage of his position as congress organizer at the closing ceremony to push the "Nice scheme" for congresses to have no short verbal communications. Addressing the assembly of the congress, he asked those who were in favor of the new scheme "to raise their hands." The poll showed that two thirds of the congress were in favor of abolishing short talks. Fortunately (in my opinion), that scheme has not been used since. It is true that the "Nice scheme" has the advantage of a large number of specialized lectures. On the other hand, a look at the lists of short talks delivered at any congress shows that it has traditionally been the entrance for many young mathematicians to the international congresses. The ICM was conceived to be different from simple symposia of specialists; this is a fact that should not be forgotten.

Cartan in his closing speech announced the election of Komaravolu Chandrasekharan as president of the International Mathematical Union for the period 1971–1974. He also said:

> We should all rejoice because this congress has allowed delegations from almost all countries where mathematics is cultivated to gather in Nice. It would have been desirable that the participation of some be more complete; I hope this will be the case for the 1974 Congress.

In this somewhat subtle and enigmatic phrase, Cartan was referring to certain absences among the Soviet delegation; particularly, that of the Fields medalist Sergei Novikov. The cause of his absence was simple: the Soviet authorities did not allow him to travel to the congress. And his was not the only case: 22 other Soviet mathematicians who had been invited to the congress were not allowed to come, among them: Dynkin, Gelfand, Gromov, Kajdan, Linnik, Manin, Moishezon, Shaferevich, Sinai.

In any case, Novikov was able to receive his medal a year later. On the occasion of a visit of Cartan to Moscow to attend the congress honoring Vinogradov's 80th birthday and an IMU meeting, a dinner was organized where Novikov received the medal from Cartan.

The congress ended with the invitation issued by H. A. Heilbronn on behalf of the Société Mathématique du Canada and the Université de la Colombie Britannique to hold the 1974 congress in Vancouver.

VANCOUVER 1974

The present generation has been engulfed by a wave of anti-intellectualism, with the result that most universities are short of students. Young people find that the problem of looking for a job is not facilitated by a university education. The idea of "art for art's sake" is less prevalent than it used to be.

THIS PESSIMISTIC PANORAMA was described by Harold Scott MacDonald Coxeter, the renowned geometer from the University of Toronto (and president of the 1974 congress), at the opening ceremony of the congress, held in the Queen Elizabeth Theatre of Vancouver on the morning of August 21, 1974.

Coxeter probably was reflecting the feeling of a whole generation—he was 67 years old at that time. The rise in the late 1960s and early 1970s of the so-called counterculture movement had a deep impact on all aspects of society. New standards of behavior such as the freedom from conventions (family, respect for authority, and even norms of sexual behavior), the change of dress codes (suits and ties were abandoned for blue jeans and T-shirts), and political activism (pacifism, civil rights, and environmental issues) became important to a new generation. Universities and academic life, once a bastion of rigor and single-minded focus, sprouted the seeds of these movements, creating a generational schism. All this was somehow reflected in the 1974 international congress in Vancouver, the capital of Canada's West coast counterculture.

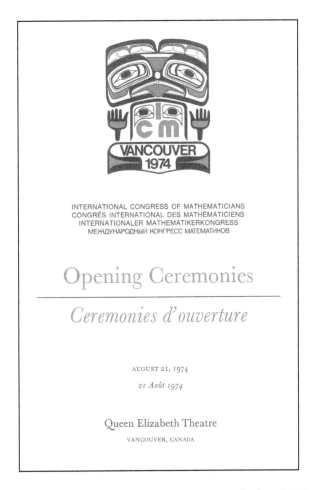

INTERNATIONAL CONGRESS OF MATHEMATICIANS
CONGRÈS INTERNATIONAL DES MATHÉMATICIENS
INTERNATIONALER MATHEMATIKERKONGRESS
МЕЖДУНАРОДНЫЙ КОНГРЕСС МАТЕМАТИКОВ

Opening Ceremonies
_____

Ceremonies d'ouverture

AUGUST 21, 1974
21 Août 1974

Queen Elizabeth Theatre
VANCOUVER, CANADA

Program for the opening ceremony of the 1974 congress. (Courtesy of the Canadian Mathematical Society, formerly Canadian Mathematical Congress.)

Vancouver, venue of the 1974 congress. (Courtesy of the University of British Columbia Archives.)

The University of British Columbia, host of the 1974 congress. (Courtesy of the University of British Columbia Archives.)

Centre d'accueil Congrès International des Mathématiciens 1974. (Courtesy of the University of British Columbia Archives.)

Discussion after a lecture at the 1974 congress. (Courtesy of the University of British Columbia Archives.)

Relaxed social gatherings in 1970s' style in Vancouver, at the 1974 congress. (Courtesy of the University of British Columbia Archives.)

The venue of the congress was the University of British Columbia. Its campus is located on a wonderful peninsula west of Vancouver, surrounded by the impressive scenery of the Strait of Georgia. The lectures were held in the buildings of the campus, and accommodation was arranged in the university dormitories. This created an atmosphere of comradeship, which, together with the general counterculture mood of the city, gave the congress a special tone. The photographs from the congress reflect a relaxed atmosphere, whether in the reception office, in the lecture rooms, or just when relaxing. One might interpret the image of David Mumford, dressed in a formal suit receiving the Fields Medal, as showing some unease with this formality.

Following the tradition, the main feature of the opening ceremony was the awarding of the Fields Medals. The committee for the 1974 medals consisted of J. F. Adams, K. Kodaira, L. S. Pontryagin, B. Malgrange, A. Mostowski, J. Tate, A. Zygmund, and the president of the International Mathematical Union, K. Chandrasekharan, acting as chairman. The surprise came when the awardees were announced, not because of their names, but because of the number of them. Even though it is was possible to award up to four Fields Medals, the committee "elected finally to select two names." They were

* Enrico Bombieri from the Università di Pisa,
* David Mumford from Harvard University.

Since the patron of the congress, His Excellency the Right Honourable Governor General of Canada, was not present at the ceremony, the medals were presented to Bombieri and Mumford by the Lieutenant Governor of British Columbia.

The Lieutenant Governor of British Columbia presenting the 1974 Fields Medals. (Courtesy of the University of British Columbia Archives.)

That evening came the laudations of the work of the medalists. Chandrasekharan reported on Bombieri's achievements in number theory, univalent functions, several complex variables, partial differential equations, and algebraic geometry, concluding that "Bombieri's versatility and strength have combined to create many original patterns of ideas." Tate referred to Mumford's work as a "tremendously successful multi-pronged attack on problems of the existence and structure of varieties of moduli."

The handling of the Fields Medals has always been a delicate issue. Their monetary value is large, although not overwhelming, but the symbolic value is inestimable. In any case, they are a solid piece of gold that someone has to carry. The letter from Dieudonné to Frostman regarding possible problems at French customs when carrying the medals into France for the Nice 1970 congress shows the type of difficulties that may be encountered. At the Vancouver congress, there was a "medal incident," although not noticed by the participants. Olli Lehto tells the story, which comes directly from Maurice Sion, the main organizer of the congress.

In the middle of the opening ceremony,

an un-programmed break was announced .... No one had remembered to take the Fields Medals from the safety deposit box at the bank where they were being kept. The bank was not near the Queen Elizabeth Theatre, where the opening ceremonies were taking place, and in spite of the help of a police escort, it took some time to pick up the medals.

After the "Bourbaki experience" at the Nice 1970 congress, the scientific program was completely under the control of the International Mathematical Union, so its structure went back to that of the Moscow congress: general expository addresses by invitation, 17; invited speakers lecturing within the sections, 157 (here there was a compromise between the 53 in Moscow and the 242 in Nice); 20 sections consecutively ordered; and 565 communications that were presented at the congress and their titles recorded in the proceedings.

Designating the invited speakers for any congress is always a difficult task, in which many tensions and strong disagreements occur, and the decisions taken can easily be controversial. If the Cold War happens to be one of the ingredients of the process, then the task can even become unpleasant. This is what happened with the selection of the invited speakers for the Vancouver congress. The Soviet representative in the committee appointed ad hoc by the International Mathematical Union for this aim, S. V. Jablonskii, considered that Soviet scientific institutions were in a better position than any other to judge, and hence chose those Soviet mathematicians who should be invited to the congress. He went on further and stated that:

> Several mathematicians from the Soviet Union who have no serious scientific achievements or were invited speakers at the preceding congress ... were selected ... [for] the Vancouver congress. From the other side, the mathematicians with new, interesting results, strongly recommended by the leading Soviet specialists, were not included.

This attitude was backed by the Soviet mathematician Lev Pontryagin, who was a member of the Executive Committee of the International Mathematical Union. However, the program was maintained as originally designed. The consequence was that of the 42 Soviet mathematicians invited to the Vancouver congress, only 22 attended.

VANCOUVER, CANADA · AUGUST 21-29, 1974

INTERNATIONAL CONGRESS OF MATHEMATICIANS
CONGRÈS INTERNATIONAL DES MATHÉMATICIENS
INTERNATIONALER MATHEMATIKERKONGRESS
МЕЖДУНАРОДНЫЙ КОНГРЕСС МАТЕМАТИКОВ

A poster for the 1974 congress, in four languages. (Courtesy of the Canadian Mathematical Society, formerly Canadian Mathematical Congress.)

After all the controversy, the expository addresses were

* "Critical Points of Smooth Functions," by V. I. Arnold;

* "Aspects of Modern Potential Theory," by H. Bauer;

* "Variational Problems and Elliptic Equations," by E. Bombieri;

* "Four Aspects of the Mathematical Theory of Economic Equilibrium," by G. Debreu;

* *"Poids dans le cohomologie des variétés algébriques,"* by P. Deligne;

* "Mathematical Problems of Tidal Energy," by G. F. D. Duff;

* "Recent Progress in Classical Fourier Analysis," by C. Fefferman;

* "Analysis over Infinite-dimensional Spaces and Applications to Quantum Field Theory," by J. Glimm;

* "Initial Boundary Value Problems for Hyperbolic Partial Differential Equations," by H.-O. Kreiss;

* *"Sur la théorie du controle,"* by J.-L. Lions;

* "Transversal Theory," by E. C. Milner;

* "Higher Algebraic K-theory," by D. Quillen;

* "Applications of Thue's Method in Various Branches of Number Theory," by W. M. Schmidt;

* "Eigenvalues of the Laplacian and Invariants of Manifolds," by I. M. Singer;

* "Inside and Outside Manifolds," by D. Sullivan;

* "On Buildings and Their Applications," by J. Tits;

* "Coding of Signals with Finite Spectrum and Sound Recording Problems," by A. G. Vitushkin.

As to the scientific sections, when they are compared with those of the Moscow congress, we see that there had been a process of refinement achieved by further subdivision and/or grouping of the previous sections. There were also some additions, such as the section on discrete mathematics and theory of computation and the section on applied statistics and mathematics in the social and biological sciences.

The full list of the sections was as follows (in parentheses, the number of invited lectures in the section):

* Section 1: Mathematical Logic and the Foundations of Mathematics (7),

* Section 2: Algebra (13),

* Section 3: Number Theory (7),

* Section 4: Algebraic Geometry (7),

* Section 5: Algebraic Groups and Discrete Subgroups (8),

* Section 6: Geometry (5),

* Section 7: Algebraic and Differential Topology (8),

* Section 8: Differential Geometry and Analysis on Manifolds (7),

* Section 9: General Topology and Real and Functional Analysis (8),

* Section 10: Operator Algebras, Harmonic Analysis, and Representation of Groups (8),

* Section 11: Probability and Mathematical Statistics, Potential, Measure, and Integration (10),

* Section 12: Complex Analysis (6),

* Section 13: Partial Differential Equations (10),

* Section 14: Ordinary Differential Equations and Dynamic Systems (7),

* Section 15: Control Theory and Related Optimization Problems (7),

* Section 16: Mathematical Physics and Mechanics (10),

* Section 17: Numerical Mathematics (8),

* Section 18: Discrete Mathematics and Theory of Computation (10),

* Section 19: Applied Statistics and Mathematics in the Social and Biological Sciences (7),

* Section 20: History and Education (4).

Membership by Countries

Algeria	2	Kuwait	4
Argentina	3	Lebanon	1
Australia	44	Libya	1
Austria	4	Malaysia	2
Belgium	14	Mexico	20
Brazil	8	Netherlands	39
Bulgaria	5	New Zealand	15
Canada	514	Niger	1
Chile	9	Nigeria	16
China, Republic of	11	Norway	16
Columbia	1	Pakistan	7
Congo	1	Philippines	2
Costa Rica	1	Poland	18
Cuba	1	Portugal	3
Czechoslovakia	4	Puerto Rico	2
Denmark	19	Rhodesia	1
Egypt	2	Romania	5
England	181	Saudi Arabia	1
Finland	11	Scotland	27
France	245	Senegal	10
Germany, Democratic Republic of	11	Sierra Leone	1
		Singapore	2
Germany, Federal Republic of	146	South Africa, Republic of	9
Greece	7	Spain	11
Guatemala	2	Sweden	17
Hong Kong	3	Switzerland	28
Hungary	24	Tunisia	3
India	26	Turkey	2
Iran	11	Uganda	1
Iraq	4	Uruguay	1
Ireland, Northern	1	U. S. A.	1274
Ireland, Republic of	7	U. S. S. R.	50
Israel	20	Venezuela	2
Italy	45	Vietnam, Democratic Republic	4
Ivory Coast	2	Wales	6
Japan	114	Yugoslavia	10
Kenya	3	Zaire, Republic of	3

Membership by country in the 1974 congress. (From the proceedings of the 1974 ICM, Canadian Mathematical Congress 1975.)

The donors for this congress were the usual: the Canadian Department of the Secretary of State, the universities of the British Columbia province, private donors (BC Forest Products, Imperial Oil, House of Seagrams, Trans Mountain Pipe Line Company, Hiram Walker, Warnock Hersey International, etc.), and scientific institutions, the National Research Council of Canada and the Canadian Mathematical Congress (which was the Canadian mathematical society). There was also financial collaboration coming from the United States, in particular, from the

Exhibition of portraits of mathematicians at the 1974 congress. (Courtesy of the University of British Columbia Archives.)

National Science Foundation and the American Mathematical Society. This is more than reasonable in view of the fact that of the 3120 registered members of the congress, 40 percent, 1271, were from the U.S.

The closing of the congress took place in the Frederick Wood Theatre of the University of British Columbia (see page 165) on the afternoon of August 29. The name of the president of the International Mathematical Union for the period 1975–1978 was announced: Deane Montgomery from Princeton; and Rolf Nevanlinna invited the congress to hold its next meeting in Helsinki.

In addition to the scientific accomplishments, the congress had its small role in spreading the new behavior revolution around the world, especially within the mathematical community. At the westernmost peak of the peninsula, where the university is located, is Wreck Beach. It is a beautiful beach on the waters of the Strait of Georgia, known for being one of the most popular nudist beaches in the world. Wreck Beach is adjacent to the university campus. The weather that late August in Vancouver was sunny and reasonably warm. It is remembered (although not recorded in the proceedings) how this beach was frequented by many congress attendees, not only those willing to go nude on the beach (which was nudist optional), but also by others from countries that were still ages from approaching any behavior revolution.

HELSINKI 1978

A STRONG FINNISH THEME was the dominant note at the opening of the 1978 International Congress of Mathematicians. First, because of the choice of the venue, the Finlandia Hall (the Finlandia-talo), a spectacular building in Helsinki formed of white geometric bodies, designed in the early 1970s by the renowned Finish architect Alvar Aalto (see page 165). Both the architect and the building were icons of the country. Second, because of the opening itself: it started with the Helsinki Philharmonic Orchestra playing a selection of themes from *The Tempest*, a musical piece by Finland's famous composer, Jean Sibelius.

The chairman of the organizing committee, Olli Lehto, was elected president of the congress, and, at Lehto's proposal, Rolf Nevanlinna was elected honorary president. After the speeches of protocol, the president of the International Mathematical Union, Deane Montgomery, unveiled the Fields Medal Committee he had chaired (L. Carleson, M. Eichler, I. M. James, J. Moser, J. V. Prohorov, B. Szökefalvi-Nagy, and J. Tits) and the names of the four mathematicians awarded the medal:

- Pierre Deligne from the Institut des Hautes Études Scientifiques,
- Charles Fefferman from Princeton University,
- Gregori Aleksandrovich Margulis from the Institute for Problems in Information Transmission, Moscow,

INTERNATIONAL CONGRESS OF MATHEMATICIANS
CONGRÈS INTERNATIONAL DES MATHÉMATICIENS
INTERNATIONALER MATHEMATIKERKONGRESS
МЕЖДУНАРОДНЫЙ КОНГРЕСС МАТЕМАТИКОВ

HELSINKI FINLAND

Opening Ceremonies

15 August 1978 at 9.30
Finlandia Hall

Opening program of the 1978 congress. (Courtesy of the Archives of the International Mathematical Union at the University of Helsinki.)

- Daniel Quillen from the Massachusetts Institute of Technology.

Medals were presented by Nevanlinna to Deligne, Fefferman, and Quillen. Margulis did not attend the congress.

Rolf Nevanlinna, honorary president of the 1978 congress, presenting the Fields Medals to the three medalists present (Gregori A. Margulis did not attend the congress). (Courtesy of the ICM 1978.)

J. Jalas and Helsinki Philharmonic Orchestra

Helsinki Philharmonic Orchestra playing at the opening ceremony of the 1978 congress. (Courtesy of the ICM 1978.)

Olli Lehto, President of the congress, at work with the organizing committee of the 1978 congress. (Courtesy of the ICM 1978.)

The ceremony ended with the Philamomnic Orchestra playing Sibelius' *Finlandia* and the Finnish national anthem.

After the ceremony came the reports on the work of the award winners. N. Katz explained that "Deligne's work centers around the remarkable relations ... between the cohomological structure of algebraic varieties over the complex numbers, and the diophantine structure of algebraic varieties over finite fields;" Carleson spoke of the "vitality of classical analysis and of the great contributions of Charles Fefferman;" J. Tits explained that Margulis' work "belongs to combinatorics, differential geometry, ergodic theory, the theory of dynamical systems and the theory of discrete subgroups of real *p*-adic Lie gropus;" regarding Quillen's work, I. M. James said that his "contributions to algebra are outstanding in their inventiveness, conceptual richness, and technical virtuosity."

Olli Lehto remarked at the opening that "the official mathematical program results from international collaboration governed by detailed rules issued by the IMU." In this manner, and following tradition, there were 17 scheduled one-hour lectures, thought to be general surveys for a wide audience:

- "Quasiconformal Mappings, Teichmüller Spaces, and Kleinian Groups," by L. V. Ahlfors;
- "Commutators, Singular Integrals on Lipschitz Curves and Applications," by A. P. Calderón;
- "Von Neumann Algebras," by A. Connes;
- "Classical Statistical Mechanics as a Branch of Probability Theory," by R. Dobrushin;
- "The Topology of Manifolds and Cell-Like Maps," by R. D. Edwards;
- "The Classification of Finite Simple Groups," by D. Gorenstein;
- "Micro-local Analysis," by M. Kashiwara;
- "Control Under Incomplete Information and Differential Games," by N. N. Krasovich;
- "L-functions and Automorphic Representations," by R. P. Langlands;
- "Modular Forms and Number Theory," by Y. I. Manin;

- "Linear Operators and Integrable Hamiltonian Systems," by S. P. Novikov;

- "The Complex Geometry of the Natural World," by R. Penrose;

- "Representations of Semisimple Lie Groups," by W. Schmid;

- "Absolute Continuity and Singularity of Probability Measures in Functional Spaces," by A. N. Shiryaev;

- "Geometry and Topology in Dimension Three," by W. P. Thurston;

- "History of Mathematics: Why and How," by A. Weil;

- "The Role of Partial Differential Equations in Differential Geometry," by S.-T. Yau.

However, three of the speakers did not attend the congress, Dobrushin, Krasovich, and Manin, although Manin sent a manuscript that was read at the congress. A look at the list of plenary speakers reveals all three Fields medalists for 1982: Connes, Thurston, and Yau (at the Vancouver 1974 congress, Deligne, Fefferman, and Quillen were also plenary speakers).

It was estimated that attendance at the plenary lectures, which were delivered in Finlandia Hall, was around 50 percent. However, there was one exception: André Weil's plenary lecture "History of Mathematics: Why and How." Undoubtedly, the speakers's prestige, but surely also the appeal of the topic (which has always interested mathematicians of all ages and fields) contributed to an audience of more than 3000 people, many of whom had to view the lecture on TV screens situated in the lobbies of Finlandia Hall. Weil began his lecture by declaring:

My first point will be an obvious one. In contrast with some sciences whose whole history consists of the personal recollections of a few of our contemporaries, mathematics not only has a history, but it has a long

one, which has been written about at least since Eudemos (a pupil of Aristotle). Thus the question "Why" is perhaps superfluous, or would be better formulated as "For Whom?"

From this viewpoint, Weil quoted Leibniz:

Its use is not just that History may give everyone his due and that others may look forward to similar praise, but also that the art of discovery be promoted and its method known through illustrious examples.

This passion for the history of the subject is a peculiarity that distinguishes mathematics from other sciences.

There were 119 scheduled forty-five minute lectures divided into 19 sections. In the IMU design, these lectures were intended to be accessible, at least in part, to mathematicians with closely related interests. The list of sections was very similar in spirit to that of the previous congress, with the major change being the resurgence of a section named simply Topology. Unfortunately, 14 of these invited speakers did not attend the congress.

There were about 500 ten-minute communications whose abstracts were printed in a booklet. And for the first time, posters (40) were presented.

For a country the size of Finland, small in population (4.7 million inhabitants in 1978) but not in size (since it is comparable to Italy), organizing a congress with 3038 participants and 900 people accompanying them was quite a challenge. A good example of the logistical problems that had to be tackled was the reception at the City Hall by invitation of the City of Helsinki. Two receptions were held, because the City Hall was not large enough for all participants to attend at the same time. Colored tickets had to be issued to organize attendance (see page 203). Everyone commented on the efficiency of the organization of the congress. The financial support needed for the congress came from the Finish state, the International Mathematical Union, a large number of private companies (among then, Oy Nokia Ab), and the fees paid by participants, which was $60 U.S.

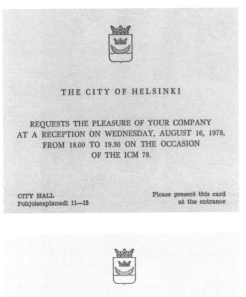

Two consecutive days and a color code were needed to hold a reception for the 4000 participants of the Helsinki 1978 congress. (Courtesy of the Archives of the International Mathematical Union at the University of Helsinki.)

The number of participating countries reached a maximum of 83 (it had been 70 in Vancouver in 1974), and the distribution by country was: U.S.A. 612, France 281, Germany (Federal Republic) 261, Finland 223, Japan 192, U.K. 173, Canada 138. The participation of the Soviet Union is a special topic that will be discussed below.

An interesting story is that of the solution given by the Polish delegation (19 mathematicians) to the high costs of attending a congress in Helsinki: it is reported that they rented a boat in which they traveled and which was used for accommodation, docking it in the harbor of Helsinki. (It was also used for enviable social gatherings). Unfortunately, we have no pictorial record of the event.

Clouds of political controversies floated over the congress. There were difficulties with the Finnish government regarding the participation of mathematicians from South Africa, due to the international stand against apartheid (just one year prior, Steve Biko had been killed while imprisoned). This was solved by the organizing committee's appeal to the principle of the free circulation of scientists. Without causing a major problem for the congress, the prudence of the organizers was able to temper a demonstration calling attention to the discrimination against the Jewish mathematicians in the U.S.S.R.

But the major difficulty was related, as in the Vancouver 1974 congress, to the choosing of invited speakers from the U.S.S.R. Pontryagin spoke about the "dissatisfaction" of the Soviet National Committee of Mathematics with the proposed lists (and also with the choosing of Fields medalists). The consequence was that Soviet attendance at the congress was extremely low: three of the 17 plenary speakers, as well as 11 of the speakers invited for forty-five minute lectures, did not attend the congress. This was also the case for the Fields medalist Gregori Margulis. Tits ended his laudation of Margulis referring to the Helsinki agreements of 1975 on international cooperation and human rights, saying:

I wish to conclude this report by a nonmathematical comment. This is probably neither the time nor the place to start a polemic. However, I cannot but express

Soviet mathematicians in the 1978 congress: A. B. Zizcenko and L. S. Pontryagin, who was blind from age 14. (Courtesy of the ICM 1978.)

my deep disappointment—no doubt shared by many people here—in the absence of Margulis from this ceremony. In view of the symbolic meaning of this city of Helsinki, I had indeed the grounds to hope that I would have a chance at last to meet a mathematician whom I know only through his work and for whom I have the greatest respect and admiration.

The final solution to the medal issue was explained by Margulis to the author (almost thirty years later):

Apparently because I could not come to the Helsinki congress, I was allowed to come to the West for the first time in 1979. It was a three month visit to Bonn which was arranged by Hirzebruch. During that visit, Jacques Tits came to Bonn and at a small ceremony presented me with the award.

At the closing ceremony, also held in Finlandia Hall, Lennart Carleson was announced as the president of the IMU for the period 1979–1982. The details of the competition for the location of the 1982 congress are known. Four countries were competing: Argentina, the Federal Republic of Germany, Israel, and Poland. The final decision was between Israel and Poland. The latter won based on its greater political stability compared to that of the Middle East. As we will soon see, this was not at all an adequate argument in favor of the Polish option, but that could not have been known at the time. These discussions were not open to the congress, since the decision was made by the IMU and then communicated to the congress.

Dinner at the boat excursion on the Gulf of Finland. Front: R. Nevanlinna and O. Lehto. (Courtesy of the ICM 1978.)

The formal invitation was issued by Kazimierz Urbanik on behalf of the Polish National Committee of Mathematics (which, formally, is the body representing national mathematicians in the International Mathematical Union):

Poland, the home of Banach, is eager to receive the world-wide mathematical community. For a long time Polish mathematicians have carried deep in their hearts the desire to organize an international congress and we are very happy that we shall now have this opportunity.

Hoping that you will accept our invitation, I welcome all of you to the next International Congress of Mathematicians to be held in August 1982 in Warsaw.

WARSAW 1982

The months of August and September encompass two important dates in the history of this country. Thirty-nine years ago on the 1 of August the Warsaw Uprising began and September 1, 1939, was the first day of the Second World War. Both these months are times of national remembrance, of reflection upon the history of our country.

During the Second World War the Polish scientific community was decimated. In particular, well over half of the actively working Polish mathematicians lost their lives. Many others found themselves in various countries all over the world. Universities, libraries and printing presses in Poland were largely destroyed. The educational system of the country was in ruins and scientific activity was disrupted.

The fact that this congress is being held in Warsaw in 1983, thirty-eight years after the war, gives evidence of the reconstruction of Polish science both in the organizational and substantive sense. In particular, it is a proof of the renaissance and expansion of the Polish mathematical community.

H ISTORIAN ALEKSANDER GIEYSZTOR gave this short but intense review of the recent tragic history of Poland and Polish mathematics. He was, at that time, president of the Polska Akademia Nauk, the Polish Academy of Sciences. The occasion was the opening ceremony of the International Congress held on August 16, 1983, in Warsaw.

General view of Warsaw in the 1980s (from the proceedings of the congress). (Courtesy of the Instytut Matematyczny Polskiej Akademii Nauk.)

Opening ceremony of ICM 1982 at the Palace of Culture of Warsaw. (Courtesy of the Instytut Matematyczny Polskiej Akademii Nauk and from author's personal files.)

The ceremony began with the men's choir of the Choral Society "Harfa" (the harp) playing the Polish national anthem. Afterward, the choir sang a selection of works:

- *Gaude Mater Poloniae*, by G. G. Gorczycki,
- *Suomen Laulu*, a Finnish song by F. Pacius,
- *Do Ojczyny*, a Polish song by F. Nowowiejski,
- *Deep River*, an American song by Purcell J. Masfield.

W. Orlicz

Władysław Orlicz (1903–1990), honorary president of the 1982 congress in Warsaw. (Courtesy of the Instytut Matematyczny Polskiej Akademii Nauk.)

Kazimier Kuratowski (1896–1980), who had always advocated holding the ICM in Poland. (Courtesy of the Instytut Matematyczny Polskiej Akademii Nauk.)

The ceremony was full of recognition for Polish mathematics and mathematicians. Czewsław Olech, president of the organizing committee and of the congress, talked about Kazimierz Kuratowski, who had advocated strongly for holding an international congress in Poland but, sadly, had died in 1980 without having seen his dream come true. Władysław Orlicz was elected honorary president of the congress and also lauded as the Nestor of Polish mathematicians, referring to the Greek character appearing in the *Iliad* as a respectable, old, and wise man, full of experience and counselor of many generations, a living example of history. There was mention of *Fundamenta Mathematicae*, the first specialized journal of mathematics, and of the Stefan Banach International Mathematical Center, created in 1972 by agreement of the academies of sciences of the socialist European countries.

The winners of international prizes were announced by Lennart Carleson, past president of the International Mathematical Union, and the awards were presented by Orlicz.

The Fields medalists were

- Alain Connes from the Institut des Hautes Études Scientifiques,
- William Thurston from Princeton University,
- Shin-Tung Yau from the Institute for Advanced Study.

The three 1982 Fields medalists and the first Nevanlinna Prize winner. (From the author's personal files.)

The committee in charge of awarding the medals was made up of H. Araki, Kyoto; N. Bogolyubov, Moscow; P. Malliavin, Paris; D. Mumford, Cambridge, MA; L. Nirenberg, New York; A. Schinzel, Warsaw; and C. T. C. Wall, Liverpool. The committee was chaired by Carleson.

Orlicz presenting the Fields Medal to Thurston. (Courtesy of the president of ICM 1982, Czewsław Olech.)

Carleson also announced to the congress the creation of the new Nevanlinna Prize on Mathematical As-

pects of Information Sciences, and the name of its first winner, Robert Tarjan from Stanford University. The committee for this prize was composed of J.-L. Lions (chairman), Paris; J. Schwartz, New York; and A. Salomaa, Turku.

The ceremony ended in the same vein as it had started, with the choir "Harfa" singing:

* *Popule meus*, by G. Palestrina,
* *Sepulto Domino*, by G. G. Gorczycki,
* *Gloria*, from *Missa Brevis* by B. Pekiel,
* *Benedictus*, by J. Fotek,
* *Sztandary* (The Banners), by W. Lachman,
* *Myszka* (A Mouse), by B. Wallek Wallek,
* *Pieśń rycerska* (Knights's Song), by S. Moniusko.

This was the first time that the names of the awardees were no surprise to the congress. What had happened? The reason was simply that the names had already been announced a year earlier in August 1982, at the General Assembly of the International

Mathematical Union held in Warsaw. The explanation for this anomaly in the established tradition was very vaguely alluded to by Carleson, who mentioned that "this congress meets under special circumstances" and more explicitly by Olech who explained that:

In April 1982, the Executive Committee of the International Mathematical Union, considering the scientific prospects of the 1982 ICM at that time, decided to postpone the Warsaw congress by one year.

We will come back to the events around the postponing of the congress after we review the scientific program.

The scientific part of the congress began with the laudations of the prize winners. H. Araki reported on the "breathtaking achievements beyond the expectations of experts" of Connes in the theory of operator algebras and the applications to differential geometry. Wall praised Thurston's "fantastic geometric insight and vision" applied to the study of topology of two and three dimensions and the interplay between analysis, topology, and geometry. Nirenberg spoke of the contributions of Yau to global differential geometry, partial differential equations, relativity theory, and algebraic geometry and said that Yau "is an analyst's geometer (or geometer's analyst) with remarkable technical power and insight." Regarding Tarjan's work on algorithm design and algorithm analysis, Schwartz said that "Tarjan has been a leader in both these ... which lie in the intellectual heart of computer science." The absences of Nirenberg and Wall necessitated the reading of their reports.

The international committee in charge of the program had chosen 16 mathematicians to deliver one-hour plenary addresses and 129 for the forty-five minute addresses in the sections. Of the invited speakers, 35 did not attend the congress, three plenary and thirty-two sectional, although five sent their manuscripts, which were read. In total, 115 invited

lectures were in one way or another delivered. Most of the canceled lectures were by Western mathematicians (a third of the U.S. invited speakers did not attend the congress). This was in some way balanced by a number of the very best Soviet mathematicians, who had not been permitted to attend other congresses but were able to come to Warsaw.

The invited plenary lectures were

"Singularities of Ray Systems," by V. I. Arnold;

"Extremal Problems in Number Theory, Combinatorics, and Geometry," by P. Erdős;

"Optimal Control of Markov Processes" by W. H. Fleming;

"Some Recent Advances in Analytical Number Theory," by C. Hooley;

"Geometric Applications of Algebraic K-Theory," by Wu-Chung Hsiang

"Problems Solved and Unsolved Concerning Linear and Nonlinear Partial Differential Equations," by P. D. Lax;

"Non-standard Characteristics in Asymptotical Problems," by V. P. Maslov;

"Modular Curves and Arithmetic," by B. Mazur;

"Global Questions in the Topology of Singular Spaces," by R. D. MacPherson;

"Structural Theory of Banach Spaces and Its Interplay with Analysis and Probability," by A. Pełczyński;

"Computational Complexity and Randomizing Algorithms," by M. Rabin;

"Turbulent Dynamical Systems," by D. Ruelle;

"Monodromy Theory and Holonomic Quantum Fields—A New Link between Mathematics and Theoretical Physics," by M. Sato;

"On Some Problems on the Continuum," by S. Shelah;

* "Some Recent Developments in Complex Differential Geometry," by Yum-Tong Siu;

* "Mathematics and Scientific Explanation," by R. Thom.

Of these lectures, the ones by Rabin, Shelah, and Thom were neither delivered nor read, because the lecturers did not attend the congress nor did they send a manuscript.

The organization of the scientific sections was very similar to that of the Helsinki 1978 congress. The main difference was the removal of Section 10 devoted to Lie groups and group representations.

All the mathematical activities of the congress took place in the Palace of Culture of Warsaw (see page 167). This enormous building was a Stalinist "gift" to the Poles, an augmented replica of the Seven Sisters skyscrapers of Moscow. Mornings were devoted to plenary lectures and afternoons to the activity of the sections. As in Moscow, the Warsaw congress had intense "elevator activity." Because of the enormous size of the building and the very large number of short communications to be presented (680; more than 800 had been submitted to the organizers), it was necessary to use 18 rooms spread throughout 12 floors of the building.

What were the "special circumstances" to which Carleson had alluded? The 1982 International Congress was scheduled to be held in Warsaw from the 11th to the 19th of August, 1982. In the autumn of 1980, a short preliminary announcement was sent "to all the countries of the world in which mathematical communities were known to exist." Again, tailoring the scientific program was a painful process, due to Soviet resistance to allowing an international committee to choose the invited speakers from the Soviet Union. This time, strong accusations were cast (anti-Soviet activity, "Zionist" propaganda), and there was a threat by the Soviet National Committee for

Mathematics to withdraw from the congress. Fortunately, the U.S.S.R. Academy of Sciences intervened, and an agreement was finally reached. By July 1981, the first announcement was sent. The second one appeared early in December 1981, including the registration form.

Simultaneously, the economic, social, and political situation in Poland was deteriorating rapidly. The Solidarność movement had started to pressure the Polish government with claims that endangered the regime. On December 13, 1981, General Wojciech Jaruzelski declared martial law. Both within and outside of Poland, Jaruzelski's primary aim, whether to save the regime or to prevent a Soviet military intervention, is still intensely debated today.

At the beginning of 1982, the question about whether to hold the congress in Warsaw was raised and discussed in the International Mathematical Union. Bad news continued: the Polish Mathematical Society was suspended, and there were mathematicians arrested and interned. A proposal from Belgium offered, as an emergency solution, to host the 1982 ICM and the General Assembly of the union. A strong controversy deeply divided the mathematical community over whether to cancel or transfer the congress to Belgium, or to keep it in Poland. What became clear was the lack of interest among mathematicians in attending the congress under the current circumstances: by spring 1982, very few registration forms had been returned.

In Poland, there was no unanimous opinion on this matter. However, the Polish organizers of the congress—and, it seems, also a substantial majority of Polish mathematicians—were very determined to hold the congress in Warsaw. More than ever, it was argued that Polish mathematicians needed contact with foreign colleagues.

At the meeting of the Executive Committee of IMU in April 1982, there was a response from the Polish Government regarding the conditions imposed

Sequence of events for the 1982 congress: the second announcement appeared in early December 1981; martial law occurs in mid-December 1981; in April 1982, the ICM is postponed to 1983. (Courtesy of the Archives of the International Mathematical Union at the University of Helsinki.)

by the union for holding the congress in Warsaw. However, there was no guarantee that the congress would be scientifically acceptable. Two decisions were taken. First, it was decided to postpone the 1982 International Congress until August 1983. The second decision was to hold the General Assembly of the union in Warsaw in August 1982, when a definite decision regarding the next ICM would be taken.

Strong controversy arose again at the General Assembly in Warsaw in August 1982. The Polish stand was still clear: ICM 1982 should be held in Poland. It was decided to delay a decision until the meeting of the Executive Committee of the union in November. Prior to the November meeting, however, there was more bad news: Solidarność was made illegal. Nonetheless, the Polish government personally assured Carleson and other officials of the International Mathematical Union that the conditions required by the union for holding the congress in Warsaw would be satisfied. These conditions were: free circulation of all scientists, reestablishment of the Polish Mathematical Society, normal conditions for transportation and communication, release of interned mathematicians. The final decision was made to hold the next international congress in August 1983 in Warsaw.

Events then moved in a positive direction: on November 15, 1982, the leader of Solidarność, Lech Wałęsa, was released. On January 1983, the third announcement of the congress was sent, with updated data; martial law was lifted in July 1983, and Polish mathematicians imprisoned for activities in support of Solidarność were released.

Not surprisingly, there was a new controversy among the international mathematical community. The basis of controversy was the same as before. Attending the congress could be interpreted as supporting the government, while not attending might be seen as abandoning the opponents to the government. There were also fears regarding personal safety. On the other hand, for mathematicians from socialist countries, whatever the circumstances were, this was a good opportunity to participate in an international congress, due to easier travel permissions and attendance conditions.

All these events had an effect on the attendance at the congress. There were 2450 participants registered (after the third announcement), although not necessarily all of them attended. The number of accompanying people was very low, 150; this reveals concerns about the political situation. The largest national group was,

Biuro Informacyjne Solidarnosci w Szwecji
Solidarity Information Office in Sweden
Solidaritets Informationskontor i Sverige

1983 - 06 - 01

TO WHOEVER MIGHT FEEL CONCERNED

The repressive Polish military junta has found its new victim:
dr Janusz Onyszkiewicz, a logician and lecturer at the department of mathematics at the Warsaw University was arrested after
his speech given at an unofficial demonstration to commemorate
the 40th anniversary of Warsaw Ghetto uprising.

Dr Onyszkiewicz, a well known Solidarity activist and its former
press secretary and spokesman, has been held prisoner ever since
under false indictment. It is but one example of constant harass-
ment Onyszkiewicz has been exposed to since he left Bialoleka in
December 1982, after having endured a year of internment there.

The World Congress of Mathematicians is going to be held in Warsaw
in August 1983...

On behalf of Solidarity Information Office in Sweden

Marek Michalski
chairman

Besöksadress	Postadress	Telefon
Västmannagatan 9 III,	111 24 Stockholm, Sweden	Phone 08/21 25 16, 08/21 55 07
Telex 19145,	Bank account PK 32052009922	Postgiro 51 34 09-3

Not everyone was happy with holding the ICM in Poland in 1983. (Courtesy of the Archives of the International Mathematical Union at the University of Helsinki.)

by far, the Poles with 820, a third of the whole congress. The next largest group were the Soviets, with 280. There were 110 participants from the U.S. and around 50 each from France and the U.K. In general, Western participation, when compared with other congresses, was reduced, and that from socialist countries was increased. The number of countries represented also decreased to 65 (it had been 83 at the previous congress).

A reason for the small number of mathematicians from the U.S. was attributed to a controversial decision of the U.S. administration, which, since January 1982, had blocked the use of federal funds to travel to Poland.

Lennart Carleson, President of the IMU, addressing the 1982 congress. (Courtesy of the president of ICM 1982, Czewsław Olech.)

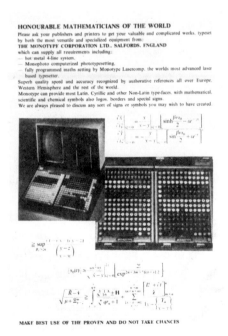

Typesetting technology in 1982 (advertisement included in the congress program). (Courtesy of the Archives of the International Mathematical Union at the University of Helsinki.)

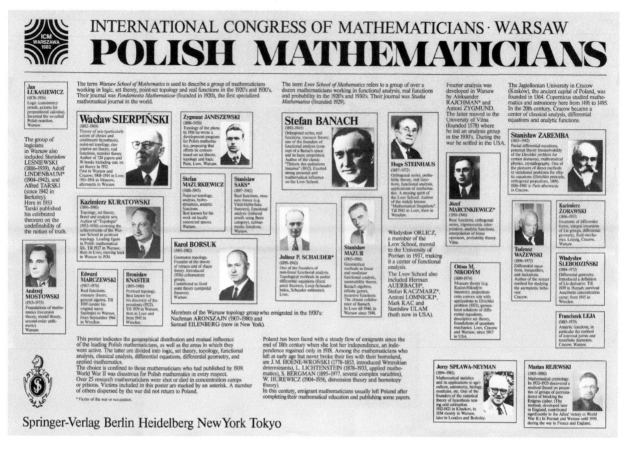

Poster featuring great Polish mathematicians. (Courtesy of Springer Science and Business Media.)

This had been decided despite recommendations to the contrary from the U.S. National Academy of Sciences.

It was quite remarkable that the Polish organizing committee was able to arrange a congress, given the precarious economic situation of the country. However, there was a reception offered by the President of the Polska Akademia Nauk in the Palace of the Council of the Ministers and even an excursion, as we will see in "Social Life at the ICM." The funds available came from the Polish Academy, from the International Mathematical Union, and from the participant fees (which were $90 U.S.).

The closing ceremony, held on August 24 in the Conference Hall of the Palace of Culture, indicated that there was a feeling of general satisfaction. The Polish mathematicians were satisfied with the outcome of the congress. Participation had been reasonably good in view of the circumstances, the scientific content had been first class, and there had been no nonscientific problems. The International Mathematical Union was relieved that international mathematical cooperation had been maintained. Those who wanted to express their solidarity with the Poles by attending the conference did so. And those who feared that their presence would be used as an endorsement of martial law and imprisonments of dissidents were satisfied by the absence of any political manipulation favoring the government. Some speakers, mainly from France, the U.K., and the U.S., were happy to have been able to publicly dedicate their lectures to Polish mathematicians who had been imprisoned.

The congress ended with the invitation by the U.S. National Academy of Sciences, presented by Jack K. Hale on behalf of the mathematical community of the U.S.A., to hold the 1986 congress at the University of California at Berkeley. It was received with acclamation.

We end with a look at the magnificent poster that Springer-Verlag issued on the occasion of the 1982 International congress (see page 216). The poster shows an intellectual landscape of classical Polish mathematics. It carries the following explanation:

> The poster indicates the geographical distribution and mutual influence of the leading Polish mathematicians, as well as the areas in which they were active.

It features 40 great Polish mathematicians (the ones who had published before 1939); for some of them a photograph is included. There is mention made of the Warsaw School of Mathematics and its journal *Fundamenta Mathematicae*, as well as the Lwów School of Mathematics and its journal *Studia Mathematicae*.

BERKELEY 1986

I accept this great honor with a good conscience because I consider myself a link between this International Congress and the one in 1936, fifty years ago, the occasion on which the Fields Medals were given for the first time.

...

At the time the circumstances were quite different; the idea of the medals had been approved in Zurich in 1932, but there had been no publicity about it and when I arrived in Oslo I did not know that the Medal had become a reality, and if I had known it I would not have considered myself the right candidate. As a matter of fact, I had not been told anything officially until I entered the room where the opening ceremony would take place, but there I was shown a place somewhere in the front, and I may have had my suspicions. Well, I had more than that. I had been warned beforehand by somebody who by mistake congratulated me a day before. But up to that point it had been a secret at least officially, even to myself. There was no tradition to go by and no protocol to follow.

IT WAS EARLY IN THE MORNING on August 3, 1986, when Lars V. Ahlfors publicly shared these reminiscences in the open-air Greek Amphitheater of the University of California at Berkeley at the opening ceremony of the International Congress. The mild weather contributed to a pleasant outdoor ceremony.

The New Albion Brass Quintet had opened the ceremony playing *Mini Overture for Brass Quintet* by

Witold Lutoslawski. Andrew Gleason was elected president of the congress, and Mary Ellen Rudin, chairman (chairwoman, as we would now say) of the U.S. National Committee for Mathematics, proposed Ahlfors for honorary president as a special celebration of the fiftieth anniversary of the Fields Medals.

Fifty years after receiving the first Fields Medal, Lars V. Ahlfors was elected honorary president of the Berkeley 1986 congress. (From the proceedings of the 1986 ICM, American Mathematical Society 1987.)

The brass quintet played again. This time it was selections from *A Brass Menagerie* by John Cheetham.

The presiding table of the 1986 congress at the opening ceremony. (From the proceedings of the 1986 ICM, American Mathematical Society 1987.)

Opening of ICM 1986 by the congress's president Andrew Gleason. (From the proceedings of the 1986 ICM, American Mathematical Society 1987.)

Next, an Associate Vice President of the university recalled that:

> Even though the founders of the City of Berkeley named it after a philosopher rather than a mathematician, Berkeley has been a hospitable environment for

mathematicians. The University is proud of its highly regarded Department of Mathematics and of the newly created Mathematical Sciences Research Institute.

Richard Johnson, Acting Science Advisor to the President of the United States, gave an interesting speech to which we will come back later, and he read a personal message from President Ronald Reagan to the congress (which ended with "God bless you").

After the brass quintet played *Fanfare* by David Amram, the ceremony of awarding the prizes commenced. Academician Ludwig Faddeev had chaired the Nevanlinna Prize Committee, including S. Cook, from the University of Toronto, and S. Winograd, from the International Business Machines Corporation, IBM. The prize was awarded to Leslie Valiant from Harvard University, who received it from Ahlfors.

Jürgen Moser, president of the International Mathematical Union and chairman of the Fields Medal Committee, introduced the rest of the members of the committee: P. Deligne, J. Glimm, L. Hörmander, K. Itô, J. Milnor, S. Novikov, and C. S. Seshadri.

THE WHITE HOUSE

WASHINGTON

July 3, 1986

I extend a warm welcome to the thousands of mathematicians
from around the globe who are attending the quadrennial
International Congress of Mathematicians. All of us are
pleased and, indeed, honored that the United States was
chosen to host this prestigious meeting.

Mathematics is the enabling force for the revolutionary
advances being made throughout the world in science and
technology. The fundamental role of mathematics is
becoming increasingly apparent in business, industry, and
government. Modern mathematicians are giving new meaning
to the famous tenet of the ancient Pythagoreans that "all
is number."

I am gratified to note that this Congress will be the
occasion for the awarding of Fields Medals to outstanding
members in recognition of their contributions to mathematics. I wish to extend my congratulations to the winners.

It is appropriate that these honors be presented at an
international meeting, for mathematics is intrinsically
international, cutting across geographical and cultural
boundaries with its own language. International competition
is a concept alien to the study of mathematics. Indeed,
cooperation between mathematicians of different nations has
been a long-standing tradition.

I wish you the best for a successful meeting. God bless
you.

Ronald Reagan's welcome letter to the 1986 congress. (From the proceedings of the 1986 ICM, American Mathematical Society 1987.)

Again, as at the Warsaw 1982 congress, not four but just three medals were awarded:

- Simon K. Donaldson from Oxford University,
- Gerd Faltings from Princeton University,
- Michael H. Freedman from the University of California at San Diego.

It was Ahlfors who presented the medals. The brass quintet closed the ceremony, playing selections from *Three Pieces for Brass Quintet*, by Minoru Fujishiro.

The congress then gathered at Zellerbach Auditorium, on the campus, for the laudations on the work of the award winners. Michael Atiyah spoke on the work of Donaldson on four-dimensional manifolds and his "result that stunned the mathematical world," which implied the existence of exotic four-dimensional spaces. Barry Mazur spoke on Faltings' proof of the half-century-old Mordell conjecture, calling it "one of the great moments in mathematics." John Milnor explained that "Freedman's 1982 proof of the 4-dimensional Poincaré hypothesis was an extraordinary tour de force." And Volker Strassen praised the multiple contributions of Valiant to the "fast growing young tree" of theoretical computer science.

All scientific sessions of the congress took place on the campus of the University of California at Berkeley. Plenary lectures were delivered in the Zellerbach Auditorium and broadcast over closed-circuit television to several large halls. Late risers were helped by technology, as the plenary lectures were videotaped and shown in the evenings.

Sixteen mathematicians were invited to deliver a plenary address:

* "Underlying Concepts in the Proof of the Bieberbach Conjecture," by L. de Branges;
* "The Geometry of 4-manifolds," by S. K. Donaldson;
* *"Neure Entwicklungen in der arithmetischen algebraischen Geometrie,"* by G. Faltings;
* "Analytical Approaches to Quantum Field Theory and Statistical Mechanics," by J. Fröhlich;
* "Topics in Quasiconformal Mappings," by F. W. Gehring;
* "Soft and Hard Symplectic Mappings," by M. Gromov;
* "Elliptic Curves and Number-Theoretic Algorithms," by H. W. Lenstra;
* "Recent Progress in Geometric Partial Differential Equations," by R. Schoen;
* "Equation Solving in Terms of Computational Complexity," by A. Schönhage;
* "Taxonomy of Universal and Other Classes," by S. Shelah;
* "Random Processes in Infinite Dimensional Spaces," by A. V. Skorohod;
* "Algorithms for Solving Equations," by S. Smale;

* "Problems in Harmonic Analysis Related to Curvature and Oscillatory Integrals," by E. M. Stein;
* "Algebraic K-theory of Fields," by A. A. Suslin;
* "Representation of Reducible Lie Groups," by D. A. Vogan;
* "Physics and Geometry," by E. Witten.

Two speakers did not attend the congress, Fröhlich and Suslin; Suslin sent his paper, which was read by E. Friedlander.

To the 19 scientific sections of the congress, 148 mathematicians were invited to deliver forty-five minute addresses. Again, only 132 were present at the congress; 11 of the missing lectures were read.

The information provided by the congress allows us to gauge the relative sizes of the mathematical groups at the time and their weight among the mathematical community. For this, we list the sections and, in parentheses, first we show the number of invited lectures and then the number of short communications presented in that section:

* Section 1: Mathematical Logic and the Foundations of Mathematics (6, 18),
* Section 2: Algebra (9, 72),
* Section 3: Number Theory (7, 39),
* Section 4: Geometry (10, 45),
* Section 5: Topology (7, 54),
* Section 6: Algebraic Geometry (8, 15),
* Section 7: Complex Analysis (9, 44),
* Section 8: Lie Groups and Representations (8, 16),
* Section 9: Real and Functional Analysis (13, 114),
* Section 10: Probability and Mathematical Statistics (7, 38),
* Section 11: Partial Differential Equations and Dynamic Systems (13, 35),
* Section 12: Ordinary Differential Equations and Dynamic Systems (11, 43),
* Section 13: Mathematical Physics (7, 46),

- Section 14: Numerical Methods and Computing (8, 33),
- Section 15: Discrete Mathematics and Combinatorics (7, 47),
- Section 16: Mathematical Aspects of Computer Science (6, 10),
- Section 17: Applications of Mathematics to Nonphysical Sciences (5, 19),
- Section 18: History of Mathematics (4, 14),
- Section 19: Teaching of Mathematics (3, 29).

The section structure was by now very stable, the main changes being in the sections related in one way or another to computers, Sections 14–16 at this congress; in 1982 in Warsaw, the sections related to computers were:

- Section 14: Control Theory and Optimization,
- Section 15: Numerical Methods,
- Section 16: Combinatorics and Mathematical Programming,
- Section 17: Computer and Information Sciences.

Overall, there were 731 ten-minute communications. The names of the authors and the titles were printed in the proceedings of the congress (an abstracts summary was distributed at the congress). Reports from the congress say that there were a "large number of really remarkably well thought-out and presented addresses."

There was a special program by the International Commission on Mathematical Instruction organized in seven sessions, and four expository lectures organized by the U.S. commission of the ICMI. Also, there was an exhibition of educational materials consisting of 40 booths, where 26 companies presented their materials.

This attention paid to educational issues was also very much present in the address of the President's Science Advisor:

We in the United States recognize that we have a serious problem. Fewer and fewer young people are studying science, mathematics, engineering, and technology; fewer and fewer people are pursuing advanced degrees in these disciplines; and fewer and fewer young people are choosing careers in scientific and technological fields.

. . .

You have the rare talent to appreciate fully the depth and power of your subject. But, with ability and talent come obligations and responsibilities, especially to those not so blessed. What you do in your laboratories, at your computer terminals, and in your studies is remarkable—indeed, it has changed the course of the world. But you must do more. It is no longer enough to be [a] practitioner of science. You must also become citizens of science—leaders of a uniquely gifted community, bound by the common goal of assuring that the progress that you have made and the improvements you have wrought for humanity will continue. Especially, you must help to assure that there is an expanding, well-qualified generation of scientists, engineers, and technologists to succeed you.

This revealed new worries and a new viewpoint of the public regarding the role of science and scientists. This concern was reflected by the administration.

As to books, the congress witnessed an important occasion: the first book on the history of the ICMs was presented, *International Mathematical Congresses: An Illustrated History 1893–1986*, by Donald J. Albers, Gerald L. Alexanderson, and Constance Reid. Originally, the book included an account of all congresses up to 1983. However, some controversy arose over the way in which the (controversial) Warsaw congress was represented. This caused the publisher to produce a revised edition, which also included the Berkeley congress. In any case, the book was a success.

The congress was, as the Harvard 1950 congress had been, very much a U.S. congress: of the 3586 attendees, 2324 were from the U.S. (that is, 65 percent) and 1262 from other countries. This American character of the congress is also seen in the fact that there were 721 participants who registered on the spot. The number of mathematicians from the U.S.S.R. was once more very small, 57. Indeed, most of the absences were from the U.S.S.R.; of the 34 Soviet mathematicians invited, 15 did not attend.

MEMBERSHIP BY COUNTRY

Algeria	3	Japan	176
Argentina	3	Jordan	1
Australia	20	Kenya	2
Austria	8	Kuwait	9
Bahrain	2	Lebanon	1
Bangladesh	1	Luxembourg	1
Belgium	20	Malaysia	3
Botswana	1	Mexico	10
Brazil	20	Netherlands	27
Bulgaria	4	New Zealand	10
Cameroon	3	Nicaragua	2
Canada	167	Nigeria	8
Chile	3	Northern Ireland	2
China-Taiwan	17	Norway	16
Colombia	2	Oman	1
Costa Rica	1	People's Republic of China	30
Cuba	2	Philippines	2
Czechoslovakia	6	Poland	17
Denmark	14	Portugal	7
Egypt	2	Romania	3
England	94	Saudi Arabia	6
Ethiopia	1	Scotland	12
Federal Republic of Germany	119	Senegal	2
Finland	19	Singapore	5
France	117	South Africa	11
German Democratic Republic	4	South Korea	6
Ghana	1	Spain	18
Greece	6	Sweden	29
Guatemala	2	Switzerland	33
Hong Kong	16	Thailand	1
Hungary	12	Trinidad	1
Iceland	6	Turkey	4
India	21	U.S.A.	2324
Indonesia	1	U.S.S.R.	57
Iran	31	U.S.V.I.	1
Iraq	4	Uruguay	2
Ireland	4	Venezuela	3
Israel	34	Vietnam	3
Italy	47	Wales	5
Ivory Coast	9	West Indies	1
Jamaica	1	Yugoslavia	10

Total 3711

Membership by country in the 1986 congress. (From the proceedings of the 1986 ICM, American Mathematical Society 1987.)

The number of countries represented was declared to be 82 (however, the different units of the United Kingdom were counted separately, so the number was at most 79). One of the new participating countries was the People's Republic of China. After China had awakened from the Cultural Revolution and after Mao's death, it had decided to join the rest of the world, including in scientific matters, such as the international congresses and the International Mathematical Union. However, the existence of the Republic of China in Taiwan caused an innumerable number of difficulties associated with the principle (or axiom?) that "there is only one China in the world" (even though there are 1200 million Chinese people!). Finally, and only after the International Mathematical Union changed its statutes to remove the adjective "national" from the title of the delegations, China entered the union, and its mathematicians could attend the international congresses. At Berkeley, there were 30 mathematicians from the People's Republic of China (in addition to 17 from China-Taiwan).

This congress had a different touch from others, a fact that was partially related to the particular organizational structure. The invitation to hold the ICM in the U.S.A. was issued by the National Academy of Sciences, who then asked the American Mathematical Society to take care of the organization. For this, a nonprofit corporation, ICM-1986, was created, with an Executive Director, a Congress Manager, and so on.

This professional management is clearly seen in the impressive organization of the financial donors for the congress. These were classified into several categories:

1. *Grand Benefactors.* National Science Foundation, Air Force Office of Scientific Research, Office of Energy Research, Office of Naval Research, Army Research Office, Alfred P. Sloan Foundation, Vaughn Foundation Fund, American Mathematical Society, the University of California, and the membership of the AMS—we will return to this last one.

2. *Benefactor.* The International Mathematical Union.

International Congress of Mathematicians 1986
August 3–11, 1986

Staff members can be identified by the red, white and blue ribbon attached to their badge. In addition, staff members who speak languages other than English will have the following colored ribbons attached to their badges:

Armenian – Peach	Hindi – Black	Russian – Tan
Bulgarian – Ivory	Italian – Wine	Serbo-Crotian – Olive Green
Chinese – Purple	Japanese – White	Spanish – Plum
Danish – Light Pink	Korean – Powder Blue	Swahili – Teal
French – Royal Blue	Norwegian – Rust	Swedish – Orange
German – Dark Green	Polish – Light Green	Turkish – Lavender
Greek – Dark Pink	Portugese – Red	Vietnamese – Medium Blue
Hebrew – Yellow		

On how to handle Babel (ICM 1986). (Courtesy of the Archives of the International Mathematical Union at the University of Helsinki.)

3. *Patrons.* Many companies such as AT&T, Exxon, GTE Laboratories Incorporated, General Motors, IBM.

4. *Sponsors.* Among them Amoco, Honeywell, the Mathematical Association of America.

3. *Donors.* Deloitte, Hermann publishers, Pergamon Press, Springer-Verlag, Texaco, and others;

5. *Contributors.* Chevron, Chrysler, American Microsystems, the Society for Industrial and Applied Mathematics, and others.

6. *Friends.* Many universities and colleges, Polaroid.

7. "Others."

There was also the revenue from the $125 fee of the congress participants.

It should be noted that 9000 members of the American Mathematical Society made personal contributions totaling $30,000. Support also came from the Army, the Navy, and the Air Force. This role was acknowledged by the Presidential Science Advisor in his address: "...science and technology increasingly are acknowledged to be essential to a nation's economic strength, industrial productivity, national security, and overall quality of life."

The closing session of the congress was held in the Zellerbach Auditorium on August 11. The president of the International Mathematical Union, Jürgen Moser, greeted Marshall H. Stone, who was present, and recalled his instrumental role in the reestablishment of the union. Moser also expressed his great disappointment that many speakers from the Soviet Union had not come to Berkeley. He then announced the name of the president of the union elected for the period 1987–1990: Ludwig Faddeev from Leningrad.

The invitation for the 1990 congress was presented by Masayoshi Nagata: "Kyoto has been the capital of Japan for about one thousand years and can show you some of the old Japanese culture."

Interlude

SOCIAL LIFE AT THE ICM

THE ZURICH 1897 CONGRESS not only started the series of the international congresses but also established a pattern that has been followed by the subsequent ones. This is clear for the scientific scheme and goals of the congresses but also applies to their social facet. We have already seen that the regulations approved in 1897 established, in the first article, the objectives of the congress. The first one was: "To foster personal relations between mathematicians of different countries."

As already noted, it is surprising that this objective of a "social nature" was placed before the "scientific objective." This was not done by chance; Ferdinand Rudio, one of the organizers of the congress, in his report at the opening session explained that:

> It suffices to consult the program or to glance at this room to agree that the congresses already would be justified even if they did not have any other goal than offering the mathematicians of all the countries of the world the occasion to talk sincerely and to share ideas. The personal relations and the progresses that directly or indirectly report to science are always one of the primary targets of any scientific meeting.

We start by reviewing the social activity of the first congress. Even though it was the shortest of all the congresses—just three days—we will see that it exhibits (almost) all the ingredients to be found in later congresses. We follow the detailed account given in the proceedings.

The day before the congress began, Sunday, August 8, the reception committee headed by Adolf Hurwitz awaited at the train station receiving mathematicians who were arriving "happily accompanied many of them by their spouses." At eight in the evening in the rehearsal hall of the Tonhalle (the concert hall), the official welcome to the guests took place, and a light meal was offered. Warm camaraderie arose between colleagues, and participants remained for a long while "enchanted by stimulating conversations and the cheerful sound of glasses." It was already midnight when the last guests left the Tonhalle.

The next day at 1 p.m., after the opening session, a banquet was held for the mathematicians and their spouses at the Pavilion of the Tonhalle. There were musical accompaniment and toasts. Enthusiasm arose, and Mittag-Leffler proposed sending a telegram to Hermite, who had not been able to attend the meeting. It was already 4 p.m. when the group boarded the steamboat *Helvetia* and sailed, accompanied by music, for an hour through the Zürichsee (Lake Zurich) to Rapperswyl, situated at the other end of the lake. There, time was spent visiting the picturesque villages, particularly Lindenhof. The return trip was eased by refreshments,

RAPPERSWYL.
Stadt am Zürich See.

RAPPERSWEIL.
Ville sur le bord du Lac de Zürich.

Zurich's congresses always include a boat trip to Rapperswyl (in 1897 it included a "Venetian Night"). (Courtesy of the Universität Bern, "Sammlung Ryhiner.")

accompanied with Veltheimer and Regensberger wines coming from the cellars of the City of Zurich. Unfortunately, the weather did not allow the participants to enjoy the "Venetian Night" that was programmed: illuminated gondolas receiving the steamboat at its arrival. The fireworks that had been scheduled also had to be canceled. However, many public and private buildings were illuminated with flares (even Neumünster Church), and from the top of the Ütliberg (the mountain dominating Zurich) greetings were sent with special lighting effects.

Tuesday evening, participants gathered at the Tonhalle and the Belvoir Park to enjoy the fireworks that had been postponed the day before.

Wednesday at 1:30 p.m., after the closing session, mathematicians and their spouses were taken by special trains to the Ütliberg, where the closing banquet was held at the Ütliberg Hotel. After a series of toasts

(Moritz Cantor dedicated his to the women present), the banquet was finished at 4 p.m. Most of the participants remained on the mountain enjoying the marvelous views of the clear sky and the snowy peaks of the Säntis, Glärnisch, and Tödi, from the Finsteraarhorn to the Diablerets.

Before continuing, let us focus on the informal welcome gathering that occurred on the evening preceding the formal opening of the congress. It gave participants the opportunity to meet and greet each other. This was of major importance in those times, since not too many mathematicians from different countries knew each other personally. Briefly we will review what happened at other congresses.

In Paris in 1900, about half of the congress participants met the evening before the opening at the Café Voltaire. In Heidelberg in 1904, the meeting was held

Simultaneous chess games by mathematician Max Euwe, Chess World Champion, in ICM 1954. (From the Archive of the Centrum voor Wiskunde en Informatica, Amsterdam.)

in the Café Imperial (then participants went to the Stadthalle—the city hall—where a reception was offered). In Rome in 1908, congress members "and many gentlemen" gathered in the Aula Magna of the university; "the meeting extended very animatedly until half past eleven." In Cambridge in 1912, the meeting was more formal: at 9:30 p.m., the president of the Cambridge Philosophical Society, Sir G. H. Darwin, received the members of the congress in the Combination Room and Hall of St. John's College. In Bologna in 1928, there was a friendly gathering of the congress participants with the members of the Unione Matematica Italiana in the Circolo di Coltura; "the animated meeting lasted beyond midnight." The Zurich 1932 congress was held under a severe economic crisis. This was the reason that the meeting took place in Claudiusstraße 21, a student residence. In 1936, the Rector of the University of Oslo offered a reception in the Aula of the university.

There are two omissions in the above list: the Strasbourg 1920 and the Toronto 1924 congresses. This could be due to lack of information. However, it should be recalled that these are precisely the two congresses held under the exclusion policy of the International Research Council.

Did these informal gatherings continue to be held after World War II? There is little information on this question. However, it seems that they did not—a pity. Probably, the size of the congresses made these meetings much more difficult to organize. The one exception was the Amsterdam 1954 congress, where, on the evening prior to the opening of the congress, there was a gathering at Natura Artis Magistra, Amsterdam's zoological garden. There, the Dutch calculating prodigy Wim Klein, alias Pascal, gave a public performance showing his abilities, while, simultaneously, his computations were verified with a Facit calculator (a renowned brand—since the beginning of the twentieth century—of manual calculating machines).

The Dutch calculating prodigy W. Klein, alias Pascal, giving a performance, in ICM 1954. (From the Archive of the Centrum voor Wiskunde en Informatica, Amsterdam.)

Checking Pascal's calculations with the manual calculating machine Facit, in ICM 1954. (From the Archive of the Centrum voor Wiskunde en Informatica, Amsterdam.)

FOSTERING PERSONAL RELATIONS

Excursions and related events have been a highlight of the social life at the international congresses. They are the occasions in which mathematicians meet and chat more freely. By reviewing some of the most relevant ones, we also engage in a visual tour through the world that the international congresses have visited.

Unfortunately, there was no immediate continuation of the Zurich 1897 excursions: the organizers of the Paris 1900 congress considered that the *Exposition Universelle* offered sufficiently attractive opportunities for congress participants. They did not plan any special event. Many participants were unhappy about this decision.

This scarcity of entertainment was compensated for during the Heidelberg 1904 congress. On Thursday, August 11, participants were taken by train to Schlier-bach and then to Ziegelhausen; the return trip back to Heidelberg went via the Neckar River. Let us follow that "delightful *Neckarfahrt*" offered by the City of Heidelberg through the report of H. W. Tyler for the *Bulletin of the American Mathematical Society*:

> After an invasion of a Gartenwirthschaft in number for whom the landlord unfortunately could not duplicate the miracle of the loaves and fishes, decorated stone-boats were taken in the evening for the romantic descent of the Neckar. Below the arches of the Carlsbrücke, each pier poured a fountain of fire into the river, while high above the town the castle burst into dazzling light with red fires, burning steadily for some fifteen minutes. In the midst of a beautiful display of fireworks from a boat on the river, the pythagorean diagram stood out brilliantly against the sky, an appropriate symbol of the nature of the meeting. The whole effect was finely spectacular.

Participants were thrilled by the spectacle and left the boats cheering the City of Heidelberg.

The Heidelberg 1904 congress had a boat excursion on the Neckar River and fireworks from the castle. (Courtesy of the Stadt Heidelberg.)

At the Rome 1908 congress, there was an excursion to Villa Adriana, and refreshments were offered at the Philosopher's Hall. (Courtesy of Erik Veldkamp.)

The day after the closing of the Rome 1908 congress, Sunday, April 12, some 600 congress members and guests were taken by special train to Tivoli. Again, it is best to follow the report of an eyewitness, in this case of C. L. E. Moore for the *Bulletin of the American Mathematical Society*:

> The first stop was at Hadrian's Villa. Carriages were ready to take those who did not care to walk from the station to the villa. On entering the ruins, we found refreshments, provided by the municipality of Tivoli, ready and waiting to be served. After spending about two hours here, we proceeded to Tivoli, where a banquet awaited us. The banquets closed with toasts in Italian, French, German and Latin. The afternoon was spent in visiting the cascades and the Villa d'Este. The returning trains arrived in Rome about 8 p.m. This was the unofficial but real close of the Congress.

The refreshments in Villa Adriana were offered at the Philosopher's Hall. The cascades belonged to Villa Gregoriana, and the access to Villa d'Este was graciously granted by the Archduke Ferdinand of Austria.

The Strasbourg 1920 congress was packed with excursions (as many as plenary lectures!): we already commented on the one to the mausoleum of the Marshal of Saxony; to Sainte-Odile; down the Rhine River by boat visiting the ports of Strasbourg and Kehl; to Saverne. The last congress day (two days after the closing session) was devoted to an excursion to Linge. This matches well with the postwar tone of the congress: Linge was one of the bloodiest battlefields of the war. It is no surprise to realize that the organizers of the excursions were two military men, General Fetterand and Colonel Holtzapfell.

The Bologna 1928 congress offered participants the choice of three different excursions. One was to Ferrara, where Copernicus had studied canon law; another one was to Riva di Garda, the Lago di Ledro, and the Ponale. This included visiting the hydroelectric power complex (even going down to the turbine rooms!); the most popular one was the excursion to

Ravenna, which 400 people attended. The proceedings give the following account:

The conference guests were offered with seignorial generosity a vermouth to get the proceedings underway; then they set off sightseeing.

A long convoy of vehicles transported the visitors to the San Vitale pine forest where, in a clearing amidst the age-old trees, tables had been set for a lavish and sumptuous lunch given by the Mayor of Ravenna.

The service was faultless, the hospitality congenial.

After the feast, the convoy of coaches took the conference guests to Porto Corsini, where they stopped briefly to watch the tranquil sunset over the shimmering and majestic Adriatic Sea.

This is the first social event of an international congress of which we have a photographic record. Regarding this very famous photograph, George Pólya commented: "And this is Hadamard on the beach near Ravenna. If you look carefully you can see his stripped underwear. He took off his shoes and waded into the sea. Note he is still wearing his hat."

Jacques Hadamard on the beach during the excursion to Ravenna, at the Bologna 1928 congress. (Courtesy of G. L. Alexanderson.)

G. H. Hardy and J. Hadamard during the 1932 lake excursion. (Courtesy of G. L. Alexanderson.)

The Zurich 1932 congress proved Pólya's statement that "whenever a congress is in a city on the coast or on a lake, there has to be an excursion on the water." On Tuesday, September 6, there was a boat excursion to the island of Ufenau, in Lake Zurich. The trip included an invitation to the official delegates to the congress for a special tea party in the castle of Herr and Frau von Schulthess-Bodmer at Au. A tea in Rapperswyl was offered to the rest of the participants.

Because of a famous picture, the excursion on the Oslo fjord at the 1936 congress is very well known (see page 235). Some 700 persons attended. We follow that marvelous evening with the report of Waldo Dunnington:

> Thursday afternoon and evening, July 16th, were devoted to a trip through the Oslo fjord on the SS. Stavangerfjord, the largest vessel of the Norwegian American lines. The crown Prince and Princess participated in this excursion, and at 6 o'clock a banquet was served in four dining halls of the ship. The brief addresses were transmitted to all the dining halls by loudspeaker,

and were interspersed with music of Grieg, Sibelius, Strauss, and a march dedicated to Crown Prince Olaf. In the evening there was dancing and card playing, happy conversation and reminiscing, with restaurant and bar on board in full swing. The ship docked in Oslo at midnight and taxis were waiting to take the guests to their hotels.

The Pólya principle of "excursions on the water" has been repeatedly applied in many congresses. In Amsterdam in 1954, there was a boat trip through the canals and harbors of the city and a day trip through the Dutch water landscape visiting the Avifauna Park. In Edinburgh in 1958, there was a full-day steamer cruise from Glasgow down to Clyde and around the island of Bute, ending at Gourock. Paul Halmos commented: "Not everyone on the Clyde excursion was from the Soviet Union" (recalling that the Edinburgh congress was the first one to have a reasonable-sized Soviet delegation). In Stockholm in 1962, the excursion was by steamer to the outer and inner Archipelago of Stockholm, visiting Sandhamn, Utö, and Vaxholm.

Dinner at the Stavangerfjord paquebot on the excursion to Oslo's fjord, at the 1936 congress. Facing the camera: Wiener, Weyl, Fréchet, Carathéodory. (Courtesy of G. L. Alexanderson.)

Élie Cartan on the excursion on Oslo's fjord, at the 1936 congress. (Courtesy of G. L. Alexanderson.)

The excursion on the Gulf of Finland at the 1978 congress. (Courtesy of the ICM 1978.)

In Helsinki in 1978, the water excursion was a four-hour cruise in the Gulf of Finland, for which two large passengers ships were needed two accommodate 1500 people (the alternative excursion was by coach to Turku, the old capital of Finland).

The main excursion at the Moscow 1966 congress was to the Russian Golden Ring, a series of historical towns around Moscow (Suzdal, Vladimir, and others) featuring ancient churches, monasteries, and fortresses. The participants visited Leo Tolstoy's home in Yasnaya Poliana. The Polish excursion in 1983 was to Bogusławice. It included a picnic party at the State Stallion Stud and the spectacle *The Cracovian Wedding*, consisting of folk songs and dances in the colorful dresses of the Cracow region and a parade of riders and coaches pulled by horses in the traditional Cracovian harnesses.

The spectacle *The Cracovian Wedding* on the excursion to Bogusławice, at the Warsaw 1982 congress. (From author's personal files.)

Western-style barbecue and rodeo in the Cow Palace in San Francisco, at the Berkeley 1986 congress. (From the proceedings of the 1986 ICM, American Mathematical Society 1987.)

Of a different character was the excursion at the Berkeley 1986 congress to the Cow Palace, a well-known arena near San Francisco where many mass events have taken place (among them, rock concerts by Elvis Presley, the Beatles, the Rolling Stones, the Doors, etc.). There, some 2800 congress members attended a rodeo and a Western-style barbecue.

At the Kyoto 1990 congress, there was a visit to the city of Nara, with its temples and shrines. Congress members visited the Kasuga Shrine.

Visit to the Kasuga Shrine, at the 1990 congress. (Courtesy of the ICM 1990.)

ROUTE MAP OF TRANSCONTINENTAL EXCURSION. TORONTO TO VANCOUVER, AND VICTORIA. AUGUST 17 TO SEPTEMBER 3, 1924. PLACES WHERE STOPS WERE MADE ARE INDICATED IN SOLID BLACK

Map of the western transcontinental excursion, organized for the 1924 congress, from Toronto to Vancouver and Victoria to see the physical features of Canada. It lasted 18 days! (From the proceedings of the 1924 ICM, The University of Toronto Press 1928.)

Outsourcing was the innovation of the last congresses regarding excursions. Travel agencies arranged to offer a variety of excursions to congress participants. This clearly eased the organizers' work. The side effect was that excursions lost part of their charm, were less recorded, and are not so interesting for ICM recollections.

The most spectacular of all the ICM excursions was the "Western Excursion" organized by the Toronto 1924 congress. The aim was that participants could "inform themselves more fully in regard to the physical features of Canada and its natural resources." The account of the proceedings tells us that:

> On the night of August 17 a number of members of the congress left Toronto on a transcontinental excursion to Vancouver and Victoria, which had been ar-

ranged through the courtesy of the Canadian National and the Canadian Pacific Railways for overseas members of the congress and of the British Association for the Advancement of Science, whose Sessions, held also in the buildings of the University of Toronto, had in part coincided with those of the congress. The excursion returned to Toronto on September 3.

The excursion lasted 18 days! We have already mentioned that it was precisely after this excursion that John C. Fields' health first broke down.

EARTHLY DELIGHTS

With the aim of promoting personal relations among mathematicians, many social activities of the international congresses have been centered on delight. Here we focus on the earthly delights, seen in banquets, receptions, balls, and parties. By looking at these activities, we sense how congresses have changed with time: from small elite meetings to massive open gatherings. We see formal balls replaced by parties, and champagne substituted with beer. We engage in a journey through history, culture, and world geography.

The Heidelberg 1904 congress is a good example of the grand style of the social activity of the first congresses. On the second congress day, there was a banquet at the Stadthalle for some 600 members and guests. The heir of the Archduke of Baden attended the banquet representing his father (the Archduke was at the time in St. Moritz following the advice of his doctors). The next day, there was a reception, which is best followed in Tyler's account:

> The visit to the grand ducal palace at Schwetzingen was the social event on Wednesday and indeed of the entire week. The visitors were met at the station by assembled fire companies, veterans, musical societies and school children and had the distinction—unusual for mathematicians—of marching to the castle gates between these rank of honor strolls in the pleasant old park and a collation in the orangery brought this Nicht-Sitzung to a successful end.

Two days later, there was another social evening: from the Scheffel terrace the congress members enjoyed the illuminating of Heidelberg's castle and dined in the castle's restaurant at the invitation of the Deutsche Mathematiker-Vereinigung. The next day, after the closing session, some visited the astrophysical institute, and others followed the invitation of Baron von Bernus and visited his collections at Neuburg Palace.

We commented before that the first photographs that we have from a social event of an international congress are from the Bologna 1928 ICM. Some were taken at a dinner at the Littoriale. This was a complex of sports facilities built under the Fascist regime; at that time, it was presided over by a great bronze equestrian statue of Mussolini. More than 1000 people attended.

George Pólya and his wife at the lunch in the Littoriale sports complex, at the Bologna 1928 congress. (Courtesy of G. L. Alexanderson.)

Gaston Julia and Charles de la Vallée Poussin at the lunch in the Littoriale sports complex, at the Bologna 1928 congress. (Courtesy of G. L. Alexanderson.)

The 1954 congress ball at the "Bellevue." (From the Archive of the Centrum voor Wiskunde en Informatica, Amsterdam.)

The official banquet of the Amsterdam 1954 congress was held in the Wintergardens of the Grand Hotel Krasnapolsky; 1500 guests attended. But the evening party at the complex "Bellevue" is more interesting. It was a dress ball in grand style, including all classical ingredients: dress suits, orchestra, balloons, and performances (by the illusionist Driebeek and, again, by the calculating prodigy Pascal). Of a similar nature was the social evening with buffet and dancing held in the Stadshuset (city hall) of Stockholm at the 1962 congress.

Buffet and dancing at the City Hall of Stockholm, during the 1962 congress. (Courtesy of the Center for History of Science of the Royal Swedish Academy of Sciences.)

Reception at the Banquet Hall of the Kremlin Palace of Congresses, in ICM 1966. (Photographs taken at the 1966 ICM in Moscow by S. V. Smirnov, from Ivanovo State University.)

Relaxed social gatherings in '70s style in Vancouver, at the 1974 congress. (Courtesy of the University of British Columbia Archives.)

The Moscow 1966 congress had a distinct character. According to the book *International Mathematical Congresses: An Illustrated History 1893–1986*, the Moscow 1966 congress was a "heady mixture of mathematics, vodka and caviar." The grand reception of the congress was held after the closing session in the Banquet Hall of the Kremlin Palace of Congresses. There is an interesting recollection of the banquet by a Western participant: it was magnificent, the menu included caviar, vodka. Strangely, however, there were no chairs. This came from old Russian wisdom: if there are no chairs, less vodka is drunk.

The real change in these earthly activities came with the Vancouver 1974 congress. Social life at the congress, in the summer of 1974, was very relaxed. Chatting on the grass, drinking beer, and listening to live music was a marked difference from other congress. We can say that it was the "hippie ICM." (This congress showed that the ICM was able to reflect the changes occurring in the world and to keep up with the younger generation's viewpoints. This was very important for the ICM's future.)

Live music in Vancouver at the 1974 congress. (Courtesy of the University of British Columbia Archives.)

Relaxed social gatherings in '70s style in Vancouver, at the 1974 congress. (Courtesy of the University of British Columbia Archives.)

Chancellor's outdoor reception in the Faculty Glade, at the 1986 congress. (From the proceedings of the 1986 ICM, American Mathematical Society 1987.)

Buffet lunch after the opening ceremony at the International Conference Center at the 1998 congress. (Courtesy of the editors of the Proceedings of the 1998 Congress.)

Outdoor receptions were the solution for the logistical problems of the large number of participants and for the new taste of the times. At the 1986 congress, there was an outdoor reception in the Faculty Glade, hosted by the Chancellor of the University of California at Berkeley. At the 1994 congress, the open-air buffet banquet was at the Irchel Campus of the University of Zurich. The logistics were so complicated that the organizers described this reception as "the miracle of the Loaves and Fishes."

The ICM party in Beijing. (From the ICM 2002.)

The 1998 congress inaugurated the "ICM party." It was held in the Mensa (student's dining hall) of the Technische Universität, Berlin. Since then, all congresses have an outdoor party with music: Beijing in 2002 and Madrid in 2006, in the Botanical Gardens of the Universidad Complutense de Madrid.

One may get the idea that there was a lot of drinking at the international congresses. However, a detailed investigation shows that we can apply to the entire ICM series the observation by Virgil Snyder in his report on the Cambridge 1912 congress for the *Bulletin of the American Mathematical Society*: the social program consisted of receptions, recitals, excursions, "and many teas."

NON-EARTHLY DELIGHTS

Music has always accompanied the international congresses except for the first congress, despite the fact that participants visited the Tonhalle several times.

The first orchestral concert was in Rome in 1908, conducted by the maestro Mancinelli in the Amphitheatre Corea (a newly inaugurated hall in the upper part of the Mausoleum of Augustus). In Bologna in 1928, the concert consisted of a selection of works by celebrated Italian composers from the nineteenth and twentieth centuries, conducted by the maestro Guarnieri in the Municipal Theatre. The concert in Zurich in 1932 was organized at the Tonhalle as an homage to the International Congress of Mathematicians; the conductor was Volkmar Andreas, and the program consisted of *Vom Fischer un syner Fru* by Othmar Schoek (concert version of the one-act opera), *Pastorale d'été* by Arthur Honegger (for orchestra), and Beethoven's Third Symphony *Eroica*. In Amsterdam in 1954, the Concertgebouw orchestra, conducted by the eminent Eduard van Beinum, played *Second Suite in B minor* by J. S. Bach, *Concert in F minor*, KV 459 by W. A. Mozart, the *Symphonische Etude* by Hendrik An-

driessen, and *La Mer, Trois Esquisses Symphoniques* by C. Debussy. Participants in the Moscow 1966 congress were invited to attend a concert with the participation of the prize winners of the renowned International Tchaikovsky Competition. In Berkeley in 1986, there was a classical concert for 1500 congress participants.

Организационный Комитет
Международного Конгресса математиков

приглашает Вас

НА КОНЦЕРТ

с участием лауреатов
III Международного Конкурса им. Чайковского

Концерт состоится 23 августа
в Актовом зале МГУ
(Ленинские горы)

Начало в 19 часов

The Organizing Committee
of the International Congress of Mathematicians

invites you

TO A CONCERT

with participation of prize-winners
of III Chaikovsky International Contest

The concert will take place in the Assembly
Hall of Moscow University (on Lenin Hills)

on August 23 at 7 p. m.

Invitation to a concert with the participation of the prize winners of the renowned International Tchaikovsky Competition, at the Moscow 1966 congress. (Courtesy of Academician A. Ershov archive, http://ershov.iis.nsk.su/archive/eaimage.asp?lang=2&did=30577&fileid=176668.)

Piano and violin music have had a special presence. Two piano concerts were arranged in Helsinki in 1978, one by Minna Pöllären in Temppeliaukio Church and another by Andrei Gavrilov in the Finlandia Hall. In Warsaw in 1983, there was a violin recital by Aureli Błaszczok (violin) and Maria Szwajger-Kułakowska

Concert by Andrei Gavrilov, at the 1978 congress. (Courtesy of the ICM 1978.)

(piano accompaniment) in the Palace of Culture, consisting of works by F. Geminiani, J. Grahms, K. Szymnowski, H. Wieniawski, L. van Beethoven, J. S. Bach, G. Bacewicz, and C. Franck. In Zurich in 1994, there was a violin recital in the Tonhalle by Hansheinz Schneeberger; Gérard Wyss was the accompanist.

Chamber music has also been present at the congresses: the Hollands Strijkkwartet in Amsterdam in 1954; in Freemasons's Hall, in Edinburgh in 1958; in Stockholm in 1962, by the Romantic twentieth century Swedish composer, Franz Berwald, in the auditorium of the Concert Hall.

The musical program of the Stockholm 1962 congress had special features: there was a recital by the famous Swedish tenor Nicolai Gedda and a ballet in the Royal Opera House—the play *Miss Julie*, inspired by Stringberg's tragedy (a resumé of the plot was provided for participants).

National folk music has also been present. In Edinburgh in 1958, there was an evening of Scottish song and dance in the Music Hall (another option was a pro-

Detailed description of the Swedish ballet Miss Julie in the congress bulletin of ICM 1962. (Courtesy of the Archives of the International Mathematical Union at the University of Helsinki.)

SOCIAL LIFE AT THE ICM

Congress men and women joining the Finnish folklore dance at Seurasaasi island, at ICM 1978. (Courtesy of the ICM 1978.)

The opening ceremony of the 1990 congress was entertained by a *Bugaku* (court dance) entitled "Gosechi no Mai" accompanied by a *Gagaku* (court music). (Courtesy of the ICM 1990.)

gram of Scottish films at the Gateway Theatre) and an informal dance at McEwan Hall, where a team from the Scottish Country Dance Society gave a demonstration. In Stockholm in 1962, there was Swedish folk music by the Swedish composer Hugo Alfvén, in particular, *Five Biblical Paintings from Dalarna*. There was also folk dancing at Skansen (an open-air museum in the Djurgården, in Stockholm). In 1978 in Helsinki, on the island of Seurasaari, there was an open-air gathering featuring Finnish folklore. It was attended by well over 3000 people. In 1983 in Warsaw, there was a special performance by the Silesian folklore ensemble *Śląsk*. At the opening ceremony in Kyoto in 1990, there was a performance of traditional court dance, *Bugaku*, and court music, *Gagaku*. In Zurich in 1994, there was a performance by the folk music group *Trio da Besto*, together with the pantomime group *Mummemschanz*, in the Kongresssaal of the Kongresshaus.

We highlight the musical program of the Cambridge 1950 congress because of its ample and diverse character: a concert by the Busch String Quartet in Sanders Theatre; an organ recital in the Daniel L. Marsh Chapel of Boston University (the only other occasion in which there has been an organ recital was in Cambridge, England in 1912, in the Chapel of King's College); a concert of ballads of various nations by the

---

GAGAKU and BUGAKU

"Gagaku", which literally means "elegant music", was originally brought into this country from China and Korea during the age of Tenpyo (8th century A.D.), but when it attained the highest stage of popularity in the Heian period (9th to 11th century A.D.), it had been completely adapted to the Japanese style.

When Gagaku is performed with dance, it is referred as "Bugaku", which means Dance and Music. It is different in both its music and dance forms, from the Noo and Kabuki theaters which developed much later. In the Bugaku dance style the dramatic elements are of far less importance than pure dance form, and in contrast with other styles of Japanese dance, Bugaku strongly emphasizes symmetry, not only in the frequent use of paired dancers, but even in the basic movement patterns of the solo dances.

The rich instrumental color, the brilliant design of the dance costumes, and the stately, flowing quality of Gagaku music and dance create an artistic effect transcending time and culture.

ETENRAKU

The most familiar piece of Gagaku. It was brought from the Tang dynasty into Japan at the beginning of the Heian period. Exact history about this piece is not known.

GOSECHI NO MAI

It is said that when Emperor Tenmu (? —686) was playing the Koto at Yoshino detached palace in Nara, a celestial nymph came down from the sky and waved the sleeves of her kimono to the music five times.

The costume is Nyonin-Hareshozoku (ladies' gala dress) of the Heian period, popularly known as Junihitoe (twelve layered costume).

In the Tale of Genji, this dance is performed by four dancers.

It is said that the head of the court musicians had this dance put on at every ceremony where a prince becomes Emperor, and on such an occasion this dance is performed by five dancers. A new version of this dance was made to be used when the Emperor Taisho (the grandfather of the present Emperor) succeeded to the throne.

The music is played in its oldest simple form by Hyoshi (clapper), Uta (song), Wagon (six string Japanese Zither), Hichiriki (double-reed oboe-like wind musical instrument with nine holes in the bamboo tube) and Fue (flute). The piece played is "Ōuta".

Explanations of Japanese musical forms provided to congress participants. (Courtesy of the ICM 1990.)

Beijing opera, at the 2002 congress. (From the ICM 2002.)

folk-singer and guitarist Richard Dyer-Bennet in Sanders Theatre; and a concert by the soprano Helen Traubel in Symphony Hall, which "was enthusiastically received by the audience and Miss Traubel received a tremendous ovation at the end of the performance."

The only jazz concert was organized at Berkeley in 1986. Some 1500 congress members attended.

The occasions for opera have not been many. At the Paris 1900 congress, there was a gala evening at the opera; we only know that the Minister of Public Instruction and Fine Arts had reserved seats for congress members. Pergolesi's opera *Il maestro di musica* was enjoyed at the eighteenth-century Court Theatre of Drottningholm during the Stockholm 1962 congress.

Berlin's offer in 1998 was *Die Zauber Flöte* ("The Magic Flute"), by Wolfgang Amadeus Mozart, staged at the Deutsche Oper Berlin.

In 2002, congress members enjoyed selections from three Beijing operas, a traditional form of Chinese theater, at Chang'an Theatre in Beijing. The summary provided by the congress organizers explains the plots:

The Crossroads. Jiao Zan, a senior officer of the Song Dynasty, is sent into exile under guard to Shamen Island because he has killed a treacherous court official. Marshal Yang orders Ren Tanghui to protect Jia in secret. Ren and Jiao stay for the night at the Crossroad Inn. The inn-keeper Liu Lihua believes that Ren intends to murder Jiao, so he steals into Ren's room and fights with him in the dark. Only when the inn-keeper's wife comes in with a candle do the three realize in the end that all has been the result of misunderstanding.

Stealing Magic Herbs. This is an episode of The Romance of the White Snake. On the day of the Dragon Boat Festival, Xu Xian advises his wife, Bai Suzhen, to drink a medicated wine. Then he sees that Bai becomes drunk and shows herself in her true colors—a white snake. At the sight of this, Xu is scared to death. To save her husband, Bai goes to the forbidden mountains to steal magic herbs. There she fights with the guards—crane boys and deer boys—and gets the right herbs she wants.

Farewell My Concubine (The Death of Yu Ji). Liu Bang and Xiang Yu have agreed to a truce and have drawn a demarcation line at Honggou. Liu's general makes a feign surrender to Xiang and then successfully lures Xiang and his troops into an ambush. Xiang and his troops are surrounded and cannot break through. When Xiang's soldiers hear their folk songs sung by the enemy, they take it for granted that their fellow soldiers have given up fighting, and their morale goes down. Xiang realizes that the game is as good as lost and drinks to despair. He bids farewell to his lover, Yu Ji, who dances her last dance before killing herself with a sword.

Musician Tom Lehrer has several songs on mathematics. Here is one: "That's Mathematics." It was created in 1985 for a children's TV series on mathematics for U.S. public television. It was used in 1993 in the celebration at the Mathematical Science Research Institute in Berkeley in honor of Andrew Wiles' proof of Fermat's Last Theorem. It was included in The Berlin Intelligencer issued for ICM 1998. (Courtesy of Tom Lehrer © 2006.)

"Day and Night" by M. C. Escher. From the catalogue of the 1954 exhibition. (© 2008 The M. C. Escher Company-Holland. All rights reserved. www.mcescher.com.)

We conclude this section with two important cultural exhibitions organized on the occasion of international congresses. One was on Escher's work in Amsterdam in 1954, and the other one was *The Life of Numbers* in Madrid in 2006.

The Dutch graphic artist Maurits Cornelis Escher visited the Palace of the Alhambra in Granada and the mosque of Córdoba, in Spain, in the 1920s and 1930s and was fascinated by the geometric tessellations of the Medieval Islamic craftsmen. He corresponded with George Pólya—before Pólya left the ETH, and Europe, for the U.S.—regarding the classification of the symmetry groups in the plane.

In 1954, the International Congress of Mathematicians was held in Amsterdam. The organizing committee of the congress—in particular, some of its members such as N. G. de Bruijn and J. J. Seidel—had the idea of having an Escher exhibition as an adjunct to the congress, because the work of Escher "shows many mathematical tendencies and is connected in a remarkable way with the mathematical thought." The exhibition was organized at the Stedelijk Museum of Amsterdam. It was the first important exhibition of Escher's work. For many, it was the first encounter with his work.

The exhibition was opened by de Bruijn, who later wrote that:

> It was a great success. A great thing for Escher too: having it in the prestigious Stedelijk Museum, gave him recognition that he did not have before and it brought him into contact with scientists from all over the world, in particular with Coxeter and young Penrose. The effect on his work is easily seen: he had learned about the circle groups.

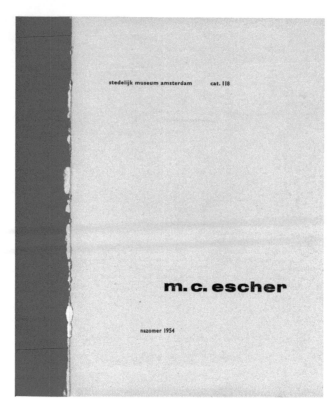

stedelijk museum amsterdam   cat. 118

m. c. escher

nazomer 1954

Catalogue of the exhibition on the graphical work of M. C. Escher organized by ICM 1954 at the Stedelijk Museum of Amsterdam. (Courtesy of the Stedelijk Museum, Amsterdam.)

In the introduction to the catalogue of the exhibition, de Bruijn attempted to explain the appeal that Escher's work has for mathematicians:

> Probably mathematicians will not only be interested in the geometrical motifs; the same playfulness which constantly appears in mathematics in general and which, to a great many mathematicians is the peculiar charm of their subject, will be a more important element.

> It will give the members of the congress a great deal of pleasure to recognize their own ideas, interpreted by quite different means than those they are accustomed to using.

In 2006, on the occasion of the international congress, the exhibition *The Life of Numbers* was presented at the Spanish National Library in Madrid. The exhibition was directed at the general public and provided an account of the relationship between human beings and numbers, from the first marks left by human hands in Palaeolithic cave paintings to the Renaissance, a journey through Mesopotamia, Egypt, Greece, Meso-America, Rome, India, and the Middle Ages. On display were Babylonian tablets, Roman coins, pre-Roman and Mayan manuscripts, an impressive collection of Renaissance mercantile arithmetics, engravings by Leonardo da Vinci and Albrecht Dürer, and maps of the Earth and the stars.

The exhibition *The Life of Numbers* was organized at the Spanish National Library by the 2006 congress. Directed at the general public, its aim was to illustrate—through manuscripts, books, and other objects drawn from the world of culture—the life course of numbers. (From the author's personal files.)

The jewel of the exhibition was the *Codex Vigilanus*, a manuscript composed in the tenth century at the Monastery of San Martín de Albelda, in Spain, currently conserved at the Monastery of El Escorial. The manuscript contains the first known written record of the full set of the Hindu-Arabic numerals.

The book *The Life of Numbers* was published for the exhibition in a beautifully illustrated edition. The introduction of the book, by Antonio J. Durán, the curator of the exhibition, reflects on the role of numbers in human culture:

Numbers may be like the gods: there is no self-respecting civilization or culture that does not include them among its achievements, although it is not clear whether such achievements should be termed intellectual, magical, or simply practical. Like the gods, numbers have different names and iconographies, depending on whether they were the product of a civilization in the South, the East, or the West. But at the same time numbers may have little in common with the gods, because, unlike the gods, numbers, although they may wear different costumes, are essentially the same, whether they be the offspring of a culture on this side of the ocean or the other, of this sea or

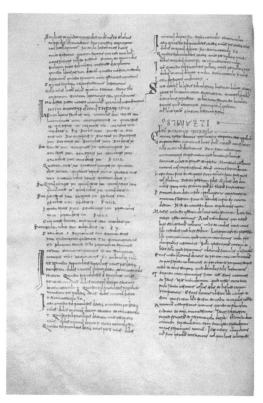

The *Codex Vigilanus*, tenth-century manuscript from the north of Spain, with the oldest record of the nine Hindu-Arabic numerals. From the exhibition *The Life of Numbers* of the 2006 congress. (© Patrimonio Nacional, Spain.)

that other even more distant one. But at the end of the day, numbers exist, which is something that perhaps cannot be said for more than one of the gods.

OLDEN TIMES

We conclude this chapter looking at a peculiar ingredient of the social activity of the ICM. Congresses have always included a social program for "associated congress members." This label nowadays comprises spouses, partners, friends—in general, anyone accompanying the congress participant. In olden times, this program was the so-called "ladies' program." Let us look at the ladies' program at the Oslo 1936 congress.

On Tuesday, July 14, there was a visit by coach to the museum on farm culture in Bigdöy and to the Viking vessels. Lunch was offered at the Dronningen Restaurant. On Wednesday morning, the National Painting and Sculpture Museum was visited; in the afternoon, there was an excursion to Frognersetteren with dinner at 7 p.m. On Friday, there was a coach excursion to Skaret. Tea was served at Mrs. C. O. Levenskiold's home. On Saturday, there was a panoramic visit by coach to Oslo and its outskirts. In appreciation for the hospitality, the foreign ladies collected a sum of money that was given to the university authorities for a student scholarship.

The world has changed a great deal since then!

PART V
IN A GLOBAL WORLD

B Y THE END OF THE TWENTIETH century, new features appear in the ICMs, marking the starting point of a new era for international cooperation in mathematics. First, the ICM opens to the East: in 1990 the congress took place in Kyoto, Japan, and in 2002 in Beijing, China. In both cases, there was an impressive response from the local mathematical community. The decision to hold the 2010 congress in Hyderabad, India, continues this trend. Second, after the isolated case of Emmy Noether in 1932, women were present at the ICM as plenary speakers; the first were Karen Uhlenbeck in Kyoto in 1990 and Ingrid Daubechies and Marina Ratner in Zurich in 1994. Third, the Berlin 1998 congress inaugurated ICM organization based on electronic communication. Finally, the role of the applications of mathematics was highlighted with the awarding, for the first time in Madrid in 2006, of the Gauss Prize.

Mathematicians looking for their place on Earth at the Beijing 2002 congress. (From the ICM 2002.)

The congresses in these period are

* Kyoto, August 21–29, 1990;
* Zurich, August 3–11, 1994;
* Berlin, August 18–27, 1998;
* Beijing, August 20–28, 2002;
* Madrid, August 22–30, 2006.

KYOTO 1990

I have often wondered why mathematicians do have congresses and what congresses mean to them. My answer is that congresses are to mathematicians what *Bon* and New Year Festivities are to Japanese, in which they abandon their daily life completely.

Japanese are believed to work continuously without vacation, but that is not true. Even in the Edo Period there were two one-week long holidays. One is the New Year Festivities and the other is the *Bon Festivities* which take place a week earlier than this time of the year.

On these holidays people are relieved from labor and go back to their native home. People are not allowed to cook on the first days of the Festivities, so that they have a busy time preparing all meals before the holiday starts.

New Year Festivities are associated with the future. We renew everything we can and start again. *Bon Festivities* are for the past. We receive ancestors' ghosts, make conversations with them, and then send them back. In cities like Kyoto people decorate their entrance halls with their treasures and keep their doors open. The whole city becomes a big museum. In the countryside people gather in the village square and dance. That is the way Japanese refresh themselves, inherit their traditions and unite. *Bon* and New Year Festivities also give young people the opportunity to meet together to make a new family.

T HESE WORDS of Hikosaburo Komatsu, president of the congress, symbolize well the Kyoto congress as a blending of tradition and innovation. They also provide an interesting interpretation of the meaning of the ICM.

Kyoto is said to be the spiritual root of Japan. (Courtesy of the ICM 1990.)

The choice of Kyoto as the first non-Western city to hold an international congress united the purity of traditional Japanese culture with the modern atmosphere of an active industrial city. On the one hand, Kyoto is

The Kyoto International Conference Hall, venue of the 1990 congress. (Courtesy of the Kyoto International Conference Hall.)

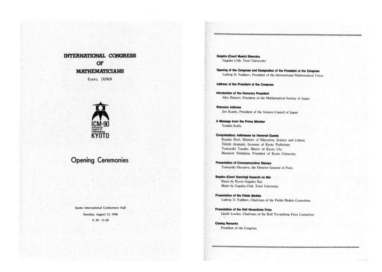

The opening ceremony program for the Kyoto 1990 congress. (Courtesy of the ICM 1990.)

▲開会式

Ludwig Faddeev, president of the IMU, opened the 1990 congress. (Courtesy of the ICM 1990.)

an ancient city, rich in cultural heritage and activities embodying the quintessence of Japanese folk art. On the other hand, in Kyoto there are companies that are technological leaders in the world. It is an academic city that has produced several world renowned scientists.

The congress took place in the impressively modern building of the Kyoto International Conference Hall.

The opening ceremony, held in the Event Hall of the Conference Center, was full of elements of Japanese culture and aesthetics. The ceremony began with the Gagaku Club of Tenri University entertaining the participants with a *Gagaku* (court music) recital entitled *Etenraku*. The president of the International Mathematical Union, Ludwig Faddeev, remarked in his opening speech on two important facts: it was the first international congress to take place outside Europe and North America, and attendance was the highest in the history of the ICM (Faddeev did not have much confidence in the offi-

cial figures of the Moscow 1966 congress). For honorary president of the congress, a pioneer of Japanese mathematics was chosen: the renowned probabilist Kiyoshi Itô.

Kiyoshi Itô, honorary president of the 1990 congress. (Courtesy of the ICM 1990.)

The 1990 Fields Medal awardees. (Courtesy of the ICM 1990.)

After the official protocol, the Kyoto Gagaku-Kai performed a *Bugaku* (court dance) entitled *Gosechi no Mai* accompanied by the Gagaku Club of Tenri University, which played another *Gagaku*.

Then came the awarding of the prizes. The Fields Medal Committee consisted of Michael F. Atiyah, Jean-Michel Bismut, Enrico Bombieri, Charles L. Fefferman, Kenkichi Iwasawa, Peter D. Lax, Igor Shafarevich, and its chairman, Ludwig D. Faddeev. Medals were awarded to

- Vladimir G. Drinfeld from the Steklov Institute in Moscow,

- Vaughan F. R. Jones from the University of California at Berkeley,

- Shigefumi Mori from the University of Kyoto,

- Edward Witten from the Institute for Advanced Study.

In the laudations, which came after the ceremony, Michio Jimbo, reading Yuri Manin's report, spoke of the "broadness, conceptual richness, technical strength and beauty of Drinfeld's work," especially on quantum groups and Galois groups of dimension one. Joan Birman recalled that Jones had "discovered an astonishing relationship between von Neumann algebras and geometric topology"; she also praised his informal style of working and his "openness and generosity ... in the best tradition and spirit of mathematics." Hironaka expressed that "the most profound and exciting development in algebraic geometry during the last decade or so was the Minimal Model Program or Mori's Program in

The 1990 Nevalinna Prize winner. (Courtesy of the ICM 1990.)

connection with the classification problems of algebraic varieties of dimension three." And Faddeev, reading Atiyah's report, referred to the "remarkable renaissance in the interaction between mathematics and physics" in which "Witten stands out clearly as the most influential and dominating figure." He noted that "in his hands physics is once again providing a rich source of inspirations and insight for mathematics."

The Rolf Nevanlinna Prize Committee was chaired by László Lovász from Budapest. It was composed of Alexandre J. Chorin from Berkeley, Michael Rabin from Jerusalem, and Volker Strassen from Konstanz. The prize was awarded to Alexander A. Razborov from the Steklov Institute in Moscow, for his "groundbreaking work on lower bound for circuit complexity."

Mathematical physics played an important role in the work of several of the Fields medalists: Drinfeld and Jones had strong connections to mathematical physics, and Witten himself was a physicist (in fact, he was the first physicist to receive the Fields Medal). Faddeev

stressed this fact at the closing ceremony; referring to the scientific program of the congress, he said, "I was glad to observe how prominently mathematical physics was represented in its connections with other domains of mathematics."

The congress had 15 plenary lectures:

- "Algebraic K-Theory, Motives, and Algebraic Cycles," by Spencer Bloch;
- "Computational Complexity of Higher Type Functions," by Stephen A. Cook;
- "Conformal Field Theory and Cohomologies of the Lie Algebra of Holomorphic Vector Fields on a Complex Curve," by Boris L. Feigin;
- "Elliptic Methods in Variational Problems," by Andreas Floer;
- "Braids, Galois Groups, and Some Arithmetic Functions," by Yasutaka Ihara;
- "Von Neumann Algebras in Mathematics and Physics," by Vaughan F. R. Jones;

The public at the opening ceremony of the 1990 congress. (Courtesy of the ICM 1990.)

The spectacular main conference room of the 1990 congress. (Courtesy of the Kyoto International Conference Hall.)

A plenary lecture at the 1990 congress. (Courtesy of the ICM 1990.)

"Geometric Algorithms and Algorithmic Geometry," by László Lovász;

"Intersection Cohomology Methods in Representation Theory," by George Lusztig;

"The Interaction on Non-Linear Analysis and Modern Applied Mathematics," by Andrew J. Majda;

"Dynamical and Ergodic Properties of Subgroup Actions on Homogeneous Spaces with Applications to Number Theory," by Gregori A. Margulis;

"Pseudodifferential Operators, Corners and Singular Limits," by Richard B. Melrose;

"Birational Classification of Algebraic Threefolds," by Shigefumi Mori;

"Hyperbolic Billiards," by Yakov G. Sinai;

"Applications of Non-Linear Analysis in Topology," by Karen Uhlenbeck;

"Multidimensional Hypergeometric Functions in Conformal Field Theory, Algebraic K–Theory, Algebraic Geometry," by Alexandre Varchenko.

The program of the congress had taken into account the proposal made by Mary Ellen Rudin, head of the U.S. delegates at the IMU General Assembly in 1986, recommending that subfields of mathematics, women mathematicians, and mathematics from small countries not be overlooked. In the list of plenary lectures, we find, for the first time since 1932, a lecture by a woman—Karen Uhlenbeck, from the University of Texas at Austin.

There were also 138 forty-five minute lectures of invited speakers delivered in the 18 scientific sections. By then, the list of the sections was almost constant, as were the numbers of invited lectures and short communications presented at each section. There was only one exception, which was the section on "Real and Functional Analysis," which had 13 invited lectures and 144 communications presented in 1986 in Berkeley, while in the Kyoto congress (here renamed "Operator Algebras and Functional Analysis") it had only six invited lectures and 52 communications. There were 620 short communications presented and more than 40 informal seminars plus other scientific meetings.

Membership by Nationality

Algeria	3	Malaysia	7
Argentina	10	Mauritania	1
Australia	21	Mexico	14
Austria	4	Mozambique	1
Bangladesh	2	Myanmar	1
Belgium	12	Nepal	1
Brazil	15	Netherlands	25
Bulgaria	8	New Zealand	7
Canada	77	Nigeria	5
Chile	4	Norway	11
China, People's Republic of	67	Pakistan	1
China-Taiwan	29	Panama	2
Colombia	1	Peru	2
Costa Rica	2	Philippines	2
Cuba	1	Poland	17
Czechoslovakia	8	Portugal	4
Denmark	15	Romania	18
Egypt	6	Saudi Arabia	2
El Salvador	1	Senegal	1
Finland	29	Sierra Leone	1
France	123	Singapore	3
German Democratic Republic	6	South Africa	11
Germany, Federal Republic of	98	Spain	24
Ghana	1	Sweden	32
Greece	8	Switzerland	14
Hong Kong	14	Tanzania	1
Hungary	15	Thailand	4
Iceland	2	Tunisia	1
India	40	Turkey	2
Indonesia	2	Uganda	2
Iran	57	United Kingdom	94
Iraq	1	Uruguay	1
Ireland	4	USA	396
Israel	30	USSR	110
Italy	69	Venezuela	2
Ivory Coast	1	Vietnam	21
Japan	2409	Yugoslavia	9
Kenya	2	Zimbabwe	2
Korea, Republic of	41	Stateless	1
Lebanon	1		
Luxembourg	1	Total	4102

Membership by nationality at the 1990 congress. (From the proceedings of the 1990 ICM, Springer 1991.)

The congress profited from what its president, Hikosaburo Komatsu, called "political reconciliation," which was the end of the Cold War due to the collapse of the Soviet Union. Attendance increased to 3954 ordinary members (plus 452 accompanying members and 92 "child members") from 76 countries. The first congress to take place in Asia showed the changes in the composition of the international mathematical com-

munity: at the 1932 congress, countries from Europe and North America comprised 83 percent of the total number, while Asia was just 4 percent; in 1990 the figures were 60 percent and 23 percent, respectively. The only dark cloud in attendance numbers, something which preluded the new problems that were to come, was that mathematicians preregistered from seven countries could not attend because of the crisis in the Persian Gulf.

The participation of Japanese mathematicians was impressive: 2409 attended the congress. This was just an example of the devotion of the Japanese mathematical community to the congress. The organization involved many people and required a large number of committees and subcommittees. The president of the organizing committee was Kunihiko Kodaira, the first Japanese Fields medalist, and the day-by-day organization was in the hands of the Research Institute for Mathematical Sciences of Kyoto University.

The second announcement for the Kyoto 1990 ICM. (Courtesy of the ICM 1990.)

The 1990 award winners before going to visit the Emperor and Empress of Japan. (Courtesy of the ICM 1990.)

The congress was sponsored by a large number of institutions and organizations, namely, the Science Council of Japan, the Mathematical Society of Japan, the Japan Society of Mathematical Education, the History of Science Society of Japan, the Institute of Actuaries of Japan, the Japan Society for Software Science and Technology, the Japan Statistical Society, the Operations Research Society of Japan, and the Information Processing Society of Japan. Of these, the Science Council of Japan is particularly interesting. It had been established after World War II, in 1949, as a government organization, formed by qualified Japanese scientists. Its aim was to promote scientific development and improve administration, industry, and living standards through science. It has had through the years an important role in the success of Japanese science and the Japanese economy.

The budget of the congress was 300,000,000 yen, of which approximately one third came from fees (30,000 yen per participant); another third came from donations by private corporations (mainly in the insurance and electronics businesses); and the rest from subventions from the International Mathematical Union, the Science Council, and the Mathematical Society of Japan (1138 individual members contributed, with more donations than the rest of the other subventions).

Surprising and unexpected was the view given by Komatsu about the attitude of the private donors:

> It is not an easy task to raise so much money. But I must confess that is was a pleasant one, too, because every executive I met for this purpose showed a liking for mathematics and appreciated that mathematics had played an important role in the development of the Japanese economy.

At the closing ceremony, the congress was informed that the Fields medalists and the winner of the Nevanlinna Prize had been invited, together with Hironaka and Itô, to visit Emperor Hirohito and the Empress in the Imperial Palace in Tokyo. The congress was also informed of the election of the president of the International Mathematical Union for the period 1991–1994, Jacques-Louis Lions, and of the site of the 1994 congress, Zurich. The ceremony was followed by a banquet with entertainment of folk music and dance.

ZURICH 1994

ZURICH HAD HOSTED the first international congress as a neutral solution to the French-German rivalry of the late nineteenth century. In 1932, choosing Zurich allowed the continuation of the delicate process of surmounting the aftermath of World War I. In 1994, Zurich was again chosen to host the ICM.

The reasons for this choice are documented in the Helsinki Archives of the International Mathematical Union. Japan had showed an interest in holding the 1986 congress, but there was the drawback of the extreme hot summer of the Japanese main islands. However, in the end Japan decided not to issue the invitation. The situation was saved by the American invitation to Berkeley. When the site for the 1990 congress was to be decided, there were two main proposals: that of Kyoto, which this time was fully prepared, and a new one presented by the Deutsche Mathematiker-Vereinigung, which proposed Munich. Despite the strong appeal of the German proposal, it was decided to favor the opening of the congresses to Asia, and hence Kyoto was chosen. It was thought that Munich would then be the site for the 1994 congress. However, the German plans were not easily reissued. Then, again the Swiss solution came to rescue the continuity of the series of international congresses.

ICM 94
International Congress of Mathematicians
Zürich 3–11 August 1994

First Announcement

zürich
m 94

Announcement for the 1994 congress. (Courtesy of the Swiss Mathematical Society.)

The opening ceremony of the 1994 congress was held at Zurich's Kongresshaus. (Courtesy of the Swiss Mathematical Society.)

A brass quintet entertained the opening ceremony of the 1994 congress. (Courtesy of the Swiss Mathematical Society.)

The 1994 award winners with Swiss Federal Minister. (Courtesy of the Swiss Mathematical Society.)

The opening ceremony took place in Zurich's Kongresshaus (the merging of the old Tonhalle—which had hosted the participants of the 1897 congress—and a modern building from the late thirties). The ceremony included many musical accompaniments: the brass quintet of Zurich Conservatory commenced with a suite from *Banchetto Musicale 1617* (*A Musical Banquet*) by Johann Hermann Schein (1586–1630).

Beno Eckmann, who had been secretary of the International Mathematical Union from 1956 to 1961, was elected honorary president and greeted the congress in three of the official languages of Switzerland (German, French, and Italian), excusing himself for not having done so in the fourth, Romansch. The quintet then played *Changing Moods* by Gordon Jacob (1895–1984). Minister Ruth Dreifuss, Head of the Federal Department of Home Affairs, gave an address with reflections on mathematics and its role in society.

The *Suite for Brass Quintet* by Edward Grieg (1843–1907) inaugurated the awarding of prizes. The Fields Medal Committee (consisting of David Mumford, as chairman, Luis Caffarelli, Masaki Kashiwara, Barry Mazur, Alexander Schrijver, Dennis Sullivan, Jacques Tits, and S. R. S. Varadhan) had chosen four mathematicians to receive the awards:

- Jean Bourgain from the IHES, the University of Illinois, and the IAS;
- Pierre-Louis Lions from the Université de Paris-Dauphine;
- Jean-Chistophe Yoccoz from the Université de Paris-Sud at Orsay;
- Efim Zelmanov from the University of Wisconsin and the University of Chicago.

The Nevanlinna Prize Committee (formed by Jacques-Louis Lions, as chairman, H. W. Lenstra, R. Tarjan, M. Yamaguti, and Y. Matiyasevich) awarded the prize to Avi Wigderson from the Hebrew University in Jerusalem.

Some lectures took place at the Kollegienhaus of Zurich University, ICM 1994. (Courtesy of the Universität Zürich.)

The awards were presented to the laureates by Beno Eckmann. The ceremony ended with the quintet playing *Trois Pastels sur la Belle Epoque* by Jean-Francois Michel.

Later, several mathematicians reported on the work of the laureates. Referring to Bourgain's work, Luis Caffarelli said that it "touches several central topics in mathematical analysis [where] he made spectacular inroads into questions where progress had been blocked for a long time." S. R. S. Varadhan said that Lions' "unique contributions . . . cover a variety of areas from probability to partial differential equations" and have always been motivated by applications. Adrien Douady said that Yoccoz is *"un pur produit, et du meilleur cru, du système français"* and explained 12 of his contributions to the theory of dynamical systems. Walter Feit explained that Zelmanov "has received a Fields Medal for the solution of the restricted Burnside problem." Yuri Matiyasevich said that Wigderson "has made a lot of wonderful contributions to diverse areas of the mathematical foundations of computer science" and high-

lighted his "impressive results connected with the so-called zero-knowledge interactive proofs."

The scientific program consisted of 16 plenary lectures, delivered in the Kongresshaus; 148 forty-five minute section lectures, delivered in auditoria of the University of Zurich and of the Eidgenössische Technische Hochschule; and almost 900 short communications presented in poster sessions. Additionally, there were five lectures by invitation of the International Commission on Mathematical Instruction and another five by invitation of the International Commission on the History of Mathematics.

The plenary lectures were

* "Transparent Proofs and Limits to Approximation," by László Babai;

* "Harmonic Analysis and Nonlinear Partial Differential Equations," by Jean Bourgain;

* "Sphere Packings, Lattices, Codes, and Greed," by John H. Conway;

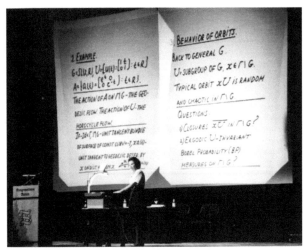

Two women were plenary lecturers at the 1994 congress. Left: Ingrid Daubechies; right: Marina Ratner. (Courtesy of the Swiss Mathematical Society.)

* "Wavelets and Other Phase Localization Methods," by Ingrid Daubechies;

* "The Fractional Quantum Hall Effect, Chern-Simons Theory and Integral Lattices," by Jürg Fröhlich;

* "Wave Propagation," by Joseph B. Keller;

* "Homological Algebra of Mirror Symmetry," by Maxim Kontsevich;

* "On Some Recent Methods for Nonlinear Partial Differential Equations," by Pierre-Louis Lions;

* "Interactions between Ergodic Theory, Lie Groups and Number Theory," by Marina Ratner;

* "Progress on the Four-Colour Theorem," by Paul Seymour;

* "Anti-self Dual Geometry," by Clifford H. Taubes;

* "Entropy Methods in Hydrodynamic Scaling," by S. R. S. Varadhan;

* "Topology of Discriminants and Their Complements," by Victor A. Vassiliev;

* "Free Probability Theory: Random Matrices and von Neumann Algebras," by Dan Voiculescu;

* "Modular Forms, Elliptic Curves and Fermat's Last Theorem," by Andrew Wiles;

* "Recent Developments in Dynamics," by Jean-Christophe Yoccoz.

Note that the number of women giving a plenary address had doubled since the Kyoto congress: there were now two, Ingrid Daubechies and Marina Ratner. In this process of breaking through the "glass ceiling," a symposium was organized, within the congress, by the Association for Women in Mathematics and the association European Women in Mathematics.

The congress also included a group of new guests: newly created countries. Henri Carnal, president of the congress, explained this in his opening address:

When we began our preparations for this event, in the summer of 1989, the borders of Europe seemed to be topologically and even metrically invariant, so that we didn't include them in the list of problems that we might have to cope with. Since then, we have witnessed the birth of many new countries and of many new mathematical societies.

4000 Mathematiker bringen 16 Millionen Franken

Wissenschafter aus 125 Ländern kommen nach Zürich zum Weltkongress

The impact of the 1994 congress in the press. (From a 1994 Swiss newspaper.)

The final collapse of the Soviet Union caused an increase in the number of countries represented in the congress, from 76 in Kyoto in 1990 to 92 in 1994. These new participants were Armenia, Azerbaijan, Belarus, Bosnia, Croatia, Czech Republic, Estonia, Georgia, Kazakhstan, Latvia, Lithuania, Macedonia, Moldova, Slovakia, Slovenia, Tajikistan, Turkmenistan, Ukraine, Uzbekistan, and Russia. Carnal had made a special mention of the Bosnian Mathematical Society, whose representatives had had to escape Sarajevo in order to attend the congress (the city was under siege and bombing since April 1992, when Bosnia and Herzegovina had declared independence from what remained of Yugoslavia).

The donors of the congress also reflected this new configuration of the world. Among others, we find the Soros Foundation and the International Association for the Promotion of Cooperation with Scientists from the Independent States of the Former Soviet Union, better known by its acronym INTAS. These groups gave financial support for 200 mathematicians from Eastern Europe.

Among the donors, there were institutions from the Swiss state, academic bodies, and private companies in insurance, banking, chemical, commerce, and industry (as, for example, the so-called Kontaktgruppe für Forschungsfragen formed by Ciba-Geigy, F. Hoffman-La Roche, Lonza, and Sandoz). The list of donors also shows new tendencies. We find IBM, which had provided financing for many of the previous congresses, but also companies exclusively devoted to software development, such as Wolfram Research Inc. In this regard, Beno Eckmann, in his address to the congress, turned around the standard saying "Whether mathematicians like it or not, the computer is here to stay" for the profounder one: "Whether the computer likes it or not, mathematics is here to stay."

J. J. Burckhardt, congress organizer of the 1932 congress, attended the 1994 congress. (Courtesy of the Swiss Mathematical Society.)

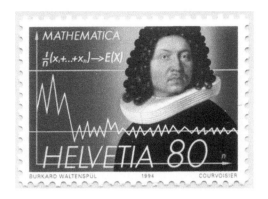

The commemorative stamp of the 1994 congress featuring Johann Bernoulli could have been prettier.

Participation was lower than expected, 2476. However, this did not ease the organizers' work, as we have seen in "Social Life at the ICM," when commenting on the open-air buffet known as "the Miracle of the Loaves and Fishes." Among the participants was J. J. Burckhardt, who 62 years before had been active in the organization of the 1932 congress.

The last plenary lecture was Andrew Wiles' on Fermat's Last Theorem. As we have already seen in "Awards of the ICM," Wiles presented his proof of the theorem in which, as he later explained, "one step in the argument was not complete." Immediately after Wiles' lecture, the closing ceremony took place. Then David Mumford was announced as the new president of the International Mathematical Union. The site of the 1998 congress was also announced: Berlin.

BERLIN 1998

In 1904 the congress was in Heidelberg, supported by Kaiser Wilhelm and the Grand Duke of Baden. This time our support comes from the Federal Republic of Germany and the Land of Berlin.

F RIEDRICH HIRZEBRUCH, honorary president of the 1998 congress, described in this manner the changes in Germany in the 94-year lapse between the two German congresses. The 1904 congress had been an exhibition of the wealth and power of the German Empire. Roman Herzog, Federal President, in his greeting message sent to the congress, expressed the meaning of the Berlin 1998 congress: "Berlin symbolizes the division of Germany, for the city itself was divided by a wall, but it also symbolizes the reunification of Germany as a democratic state with scientific freedom." Hirzebruch himself had lived the div-

View of the Reichstag under reconstruction after the Reunification (from the proceedings of the 1998 congress). (Courtesy of the editors of the Proceedings of the 1998 Congress.)

The opening ceremony of the 1998 congress took place at the International Conference Center of Berlin. (Courtesy of the International Conference Center of Berlin.)

ision very intensively: he was president of the Deutsche Mathematiker-Vereinigung in 1961 when the Berlin Wall was built and when the Mathematical Society of the DDR was created (until that moment, there had been one unified mathematical society). He was president again in 1990 when the reunification of the mathematical society occurred (following that of the country).

The opening ceremony of the 1998 congress. (Courtesy of the editors of the Proceedings of the 1998 Congress.)

Germans were happy and proud to exhibit a unified and democratic country, with restored scientific power. In Berlin, there were three universities, Freie, Technische, and Humboldt, plus another in Postdam. There were two important research institutes, the Konrad-Zuse-Center—named after the German engineer who was a pioneer in electronic computation—and the Weierstrass Institute for Applied Analysis and Stochastics.

The opening ceremony of the congress was held on August 18 in the enormous building of the International Congress Center of Berlin, with 3000 people attending and musical accompaniment by the Ensemble Oriol Berlin.

Yuri Manin, chairman of the Fields Medal Committee, conducted the ceremony of the presentations of the medals. He began recalling Georg Cantor's famous motto:

"Das Wesen der Mathematik liegt in ihrer Freiheit,"

that is, "The essence of mathematics is its freedom." The committee for the award consisted of Manin, John Ball, John Coates, J. J. Duistermaat, Michael Freedman, Jürg Fröhlich, Robert MacPherson, Kyoji Saito, and Steve Smale. Four medalists were chosen:

* Richard Ewen Borcherds from the University of California at Berkeley and the University of Cambridge,
* William Timothy Gowers from the University of Cambridge,
* Maxim Kontsevich from the Institut des Hautes Études Scientifiques,
* Curtis McMullen from Harvard University.

The medals were presented by Hirzebruch. Later on in the congress, there were reports on the work of the awardees, by Peter Goddard on Borcherds' work, by

The recipients of all the 1998 awards; in the background, Martin Grötschel, president of the congress. (Courtesy of the ICM 1998.)

Béla Bollobás on Gowers' work, by Clifford Taubes on Kontsevich's work, and by Steve Smale on McMullen's work. However, this time there was an official explanation for the awarding of the medals. Borcherds was awarded the medals for his contributions to "algebra, the theory of authomorphic forms, and mathematical physics, including . . . the proof of the Conway-Norton moonshine conjecture"; Gowers for his contribution to "functional analysis and combinatorics, developing a new version of infinite dimensional geometry, including the solution to two of Banach's problems"; Kontsevich for his contributions to "algebraic geometry, topology, and mathematical physics, including the proof of Witten's conjecture of intersection numbers in moduli spaces of stable curves"; and McMullen for his contributions to "the theory of holomorphic dynamics and geometrization of three-manifolds, including the proof of . . . Kra's theta-function conjecture."

The Nevanlinna Prize Committee had been chaired by David Mumford and consisted of Björn Engquist, Tom Leighton, and Alexander Razborov. The prize was awarded to Peter Shor from AT&T Bell Laboratories, for "having been the principal driving force behind the development of quantum computing." The award was presented by Olli Lehto on behalf of the University of Helsinki and the International Mathematical Union. Later, there was a report on Shor's work by Ronald Graham.

This time, the ceremony had a special and unique feature, corresponding to a special and unique achievement. Manin announced that the International Mathematical Union had decided to produce a commemorative silver plaque as a special tribute to Andrew Wiles, from Princeton University, "on the occasion of his sensational achievement," referring to his proof of Fermat's Last Theorem. As we have already seen in "Awards of the ICM," since Wiles was over the age limit, he could not receive the Fields Medal. Manin recalled Fermat's own stating of the problem:

"Nullam in infinitum ultra quadratum potestatem in duos ejusdem nominis fas est dividere."

Andrew Wiles received at the 1998 congress a silver plaque from the IMU in recognition for his proof of Fermat's Last Theorem. (Courtesy of the editors of the Proceedings of the 1998 Congress.)

The silver plaque was presented to Wiles by Hirzebruch. Manin remarked that "unfortunately this plaque is too small to write down Wiles' proof." The *Notices of the AMS* reported that Wiles received a "thundering round of applause longer than given to any of the other awardees." The next day, Wiles gave a special evening lecture, which attracted an audience of 2300 people. It was entitled "Twenty Years of Number Theory" and started as follows:

We begin with three problems considered by Fermat:

(1) Which prime numbers can be written as the sum of two integer squares?

(2) Is there a right-angled triangle with rational length sides and area 1?

(3) Do there exist solutions to the equation: $x^n + y^n = z^n$ for $n \geq 3$?

The answer to the first question is: there are precisely the primes p which are congruent to 1 mod 4. The answer to the second is: no. The solutions to these two problems were found by Fermat himself. The third problem of course needs no introduction.

Andrew Wiles' special evening lecture on "Twenty Years of Number Theory" had 2300 attendants. (Courtesy of the ICM 1998.)

After reviewing some of the principal developments in the field, Wiles observed that:

> One change in number theory over the last twenty years is that it has become an applied subject (Perhaps one should say it has gone back to being an applied subject as it was more than two thousand years ago).

The opening ceremony ended with the Ensemble Oriol Berlin playing a movement of Wolfgang Amadeus Mozart's *Divertimento*, Köchel 287.

The scientific program consisted of 21 plenary lectures and 169 forty-five minute invited addresses within the 19 scientific sections. There were also 1098 short fifteen-minute contributions and 236 poster presentations. Additionally, there were 235 ad-hoc talks of fifteen minutes. Thus, the total number of contributed presentations was very large: 1569. Except for the first session, which took place at the Conference Center, the sessions of the congress were held in the Technische Universität Berlin, TUB (see page 170). Plenary lectures took place in the mornings in the Audimax of the TUB, and forty-five minute lectures took place in the afternoons in six parallel sessions.

The plenary lectures were

- "Local Index Theory and Higher Analytic Torsion," by Jean-Michel Bismut;

- "Some Analogies Between Number Theory and Dynamical Systems on Foliated Spaces," by Christopher Deninger;

- "From Shuffling Cards to Walking Around the Building: An Introduction to Modern Markov Chain Theory," by Persi Diaconis;

- "Chaotic Hypothesis and Universal Large Deviations Properties," by Giovanni Gallavotti;

- "From Classical Numerical Mathematics to Scientific Computing," by Wolfgang Hackbusch;

- "Dynamics, Topology, and Holomorphic Curves," by Helmut H. W. Hofer;

The lectures of the 1998 congress took place at the Mathematics building of the Technische Universität Berlin. (Courtesy of the ICM 1998.)

- "Geometric Model Theory," by Ehud Hrushovski;
- "Constant Term Identities, Orthogonal Polynomials, and Affine Hecke Algebras," by I. G. Macdonald;
- "Applied Mathematics Meets Signal Processing," by Stéphane Mallat;
- "Fibrations in Symplectic Topology," by Dusa McDuff;
- "Solvable Lattice Models and Representation Theory of Quantum Affine Algebras," by Tetsuji Miwa;
- "Dynamical Systems—Past and Present," by Jürgen Moser;
- "Mathematical Problems in Geophysical Wave Propagation," by George Papanicolaou;
- "Operator Spaces and Similarity Problems," by Gilles Pisier;
- "L-Functions," by Peter Sarnak;
- "Quantum Computing," by Peter W. Shor;
- "The Population Dynamics of Conflict and Cooperation," by Karl Sigmund;
- "Huge Random Structures and Mean Field Models for Spin Glasses," by Michel Talagrand;
- "Geometric Physics," by Cumrun Vafa;
- "Dynamics: A Probabilistic and Geometric Perspective," by Marcelo Viana;
- "A^1-Homotopy Theory," by Vladimir Voevodsky.

The Berlin congress had some special features that distinguish it from previous congresses.

For the first time, there was an extensive use of electronic communication and information in the organization of the congress: letters were emailed, the possibility of electronic registration was offered and used by two thirds of the participants, a server with up-to-date information was set up, and 95 percent of the in-

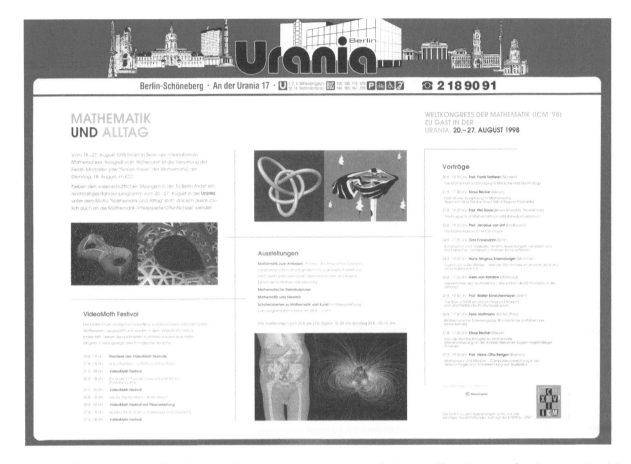

Urania, a Berlin institution with a long tradition in the popularization of science, offered lectures for the general public on mathematical topics. (Courtesy of Urania Berlin e. V.)

vited and contributed presentations were submitted electronically. The plenary and invited lectures were made available on the Internet before the congress started. The opening ceremony was transmitted worldwide on the Internet. The "electronic revolution" had reached the international congresses.

Another special feature was the plan of the organizers of extending the celebration of the International congress to the nonmathematical public. For this, an extensive series of lectures and events directed towards a general audience was presented. The program was named *Mathematik und Alltag* (Mathematics and Ev-

eryday Life). It included 11 lectures on mathematics, the VideoMath Festival (a "composition of selected short videos on mathematics"), several other mathematical films, and the exhibitions "Hands-on Mathematics" and "Mathematical Stone Sculptures." To host some of these activities, the public lecture hall Urania was rented. It is an institution with a long tradition—since 1888—in the popularization of science. It is housed in a navy blue cubic building in Berlin. The program was advertised with posters throughout the city and in the subway. Among the lecturers was, for example, the German writer and thinker Hans Magnus Enzensberger. The result was a success: more than 5000 people attended the Urania lectures and video performances, and about 10,000 visited the exhibitions in the Urania. The objective of "mathematics goes public" was achieved.

Hans Magnus Enzensberger lecturing at Urania. (Courtesy of the ICM 1998.)

Many references to the dark moments in German history permeated the congress. The Federal President Roman Herzog in his message to the congress recalled that "the Nazi regimen forced many scientists into exile or even murdered them." Along these lines, the Federal Minister of Education, Research, and Technology announced the creation of a "prize for excellent junior scientists ... named after Emmy Noether ... who had to leave Germany in 1933 without receiving the scientific recognition she would have deserved."

The most poignant words were Friedrich Hirzebruch's. When issuing the invitation for the congress in Zurich in 1994, he had explained that:

After the terrible period of World War II there were attempts to invite the congress to Germany, beginning

in the sixties; these attempts failed, always for understandable reasons.

At the opening ceremony in Berlin, Hirzebruch recalled several of the presidents of the DMV:

Alfred Pringsheim died in Zurich in 1941 at the age of 90 after having escaped from Germany. Edmund Landau lost his chair in Göttingen in 1934. Otto Blumenthal was deported to the concentration camp Theresienstadt, where he died in 1944. Hermann Weyl ... emigrated to the United States in 1933.

He also remembered Hilbert and his collaborators, recalling what Hermann Weyl wrote in the obituary of his teacher: "[when] the Nazi storm broke [they were] scattered over the earth." Hirzebruch concluded saying: "We must teach the next generation 'not to forget.'"

In this vein, there was a special session entitled *Mathematics in the Third Reich and Racial and Political Persecution*. It consisted of two talks:

* "Victims, Oppressors, Activists, and Bystanders: Scientists' Response to Racial and Political Persecution," by Joel Lebowitz from Rutgers University;

* "Mathematics and Mathematicians in Nazi Germany: History and Memory," by Herbert Mehrtens from the Technische Hochshule Braunschweig.

During this congress, the most intense of these activities was organized by the DMV to honor the memory of the victims of the Nazi terror. The official announcement read:

In 1998, the ICM returns to Germany after an intermission of 94 years. This long interval covers the darkest period in German history. Therefore, the DMV wants to honour the memory of all those who suffered under the Nazi terror. We shall do this in the form of an exhibition presenting the biographies of 53 mathematicians from Berlin who were victims of the Nazi regime between 1933 and 1945. The fate of this small group illustrates painfully well the personal sufferings and destruction of scientific and cultural life; it also sheds some light on the instruments of suppression and the mechanism of collaboration.

The exhibition *Terror and Exile* honored the memory of all those who suffered under Nazi terror, by presenting the biographies of 53 mathematicians from Berlin. Three were present at the 1998 congress. (Courtesy of the editors and the Deutsche Mathematiker-Vereinigung.)

The exhibition was entitled *Terror and Exile*. Three of those 53 mathematicians were invited to attend as guests of the DMV and of the Senate of Berlin. They were analyst Michael Golomb from Lafayette, and group theorists Walter Ledermann from London and Bernhard Neumann from Canberra. Feodor Theilheimer from the U.S.A. was represented by his daughter. The inauguration ceremony of the exhibit was very moving. The three guests gave touching addresses on their experiences of expulsion and emigration. Around 400 members of the congress attended. A 72-page pamphlet summarizing the event was published. Afterwards, Michael Golomb wrote:

> I thought is was very decent of the German mathematicians to bring up and condemn now, fifty years later, the infamous past of their country before a forum of people, many of whom did not remember or did nor know or care about these things.

Despite the success of electronic communication at the congress, books also had their place. In this case, for the second time in 12 years, a book on the international cooperation in mathematics was debuted at the congress: Olli Lehto's *Mathematics without Borders: A History of the International Mathematical Union.*

Participation in the congress consisted of 3346 mathematicians from 98 countries, and approximately 800 guests. Of the participants, 1 percent came from Australia, 2 percent from Africa, 12 percent from Asia, 20 percent from the Americas, and 65 percent from Europe. Also, 12 percent were women. There were 202 participants from the Russian Federation and 75 from Ukraine; if we add to these the participants from the other former Soviet republics (just for the sake of an intellectual exercise), we obtain 353 participants from the territories formerly part of the Soviet Union. This large number was possible thanks to a special fund of 900,000 DM and the support from the organizing committee for participants coming from Eastern Europe and the former Soviet Union. Anatoly M. Vershik, president of the St. Petersburg Mathematical Society, thanked the organizers for this effort and recalled the limited attendance of Soviet mathematicians at the international congresses in the '60s, '70s, and '80s. He noted the difficulties for the invited speakers to attend, even for the Fields medalists (referring to the cases of Novikov and Margulis).

The closing ceremony took place in the TUB. There, the newly elected president of the International Mathematical Union for the period 1999–2002, Jacob Palis, encouraged the mathematicians of the world to participate in the multiple activities of the World Mathematical Year 2000.

Martin Grötschel, president of the congress, recalled that

> At the first International Congresses it has been a tradition to commemorate the mathematicians who have deceased in the previous years. We would like to resume this tradition today. Following a German custom, I would like to ask you to stand up for a few moments and remain in silence while I read some words of remembrance.

He chose six mathematicians who had died in the last four years: Hansgeorg Jeggle, François Jaeger, André Weil, Paul Erdős, Lars Ahlfors, and Kunihiko Kodaira.

Kung Chin Chang, president of the Chinese Mathematical Society, invited the congress to meet in 2002 in Beijing, remarking that:

> The next congress, the first in the new century, will be held for the first time in a developing country. This will add a new chapter to Olli Lehto's book *Mathematics without Borders*.

BEIJING 2002

THE ATTENDANCE at the Beijing ICM was very large—the largest in the history of the ICM. It mirrored the large population of China, one fifth of the world's population, its metropolis, Beijing, with 13 million inhabitants, and its more than 2000 mathematicians in universities and research institutes. There were 4270 registered participants (even though this figure is a few less than the official one for the Moscow 1966 congress, the counting here is more reliable).

The Chinese government gave special support to the congress, as part of its plans to enhance science and technology as a base for the country's development. These plans conferred an important role to basic science, in particular to mathematics, as a symbol of modern civilization. This, coupled with the traditional Chinese respect for those who make great scientific achievements, made the ICM an important national event. The presence of the president of the People's Republic of China, Jiang Zemin, at the opening ceremony indicated this view.

The opening ceremony was held in the Great Hall of the People (see page 171), located on Tiananmen Square (which at its north end has the Gate of Heavenly Peace, entrance to the Forbidden City). The Square, capable of hosting one million people, has been the site of many crucial events in Chinese history. The ceremony took place in the Main

2002 年国际数学家大会
数学出版物及软件展览会
中国　北京　2002 年 8 月 20-28 日

The name of the congress in Chinese. (Courtesy of the ICM 2002.)

Old-fashioned official photograph of the authorities of the 2002 congress. (Courtesy of the ICM 2002.)

The opening ceremony of the 2002 congress took place at the Great Hall of the People. (Courtesy of the ICM 2002.)

Presiding table of the Beijing 2002 congress. (Courtesy of the ICM 2002.)

Auditorium of the Great Hall, which can accommodate an audience of 10,000 people—the hall where the congresses of the Chinese Communist Party take place. Participants, guests, and authorities for the opening of the ICM totaled 5000 people. (This includes a large group of children in the balcony of the auditorium wearing T-shirts with the slogan "I want to be a mathematician.")

The ceremony began with an ensemble of traditional Chinese instruments playing *Happiness* and *Blossoms and a Full Moon*. The respected Chinese mathematician Shiing-Shen Chern was elected honorary president of the congress. He was almost 91 years old. Among all the formal speeches of the ceremony, let us cite the words of the Chinese geometer Wen-Tsun Wu, who was president of the congress and, having to choose the viewpoint of a nonmathematician, cited Karl Marx's words: "Each science could be considered to be perfect only if it permits the successful application of mathematics."

There was a musical interlude with *Thunders, Colorful Clouds Chasing the Moon*, *Lake Under the Autumn Moon*, and *Every Step Going Higher*.

Next came the awarding of the prizes. The Fields Medal Committee was formed by Yakov Sinai, chairman, together with James Arthur, Spencer Bloch, Jean Bourgain, Helmut Hofer, Yasutaka Ihara, H. Blaine Lawson, Sergei Novikov, George Papanicolaou, and Efim Zelmanov. This time, the committee chose to award only two medals (as has occurred after 1966 only at the Vancouver 1974 congress). The medals were awarded to

- Laurent Lafforgue from the Institut des Hautes Études Scientifiques, for "a major advance in the Langlands Program, thereby providing new connections between number theory and analysis";

Three Presidents: Jacob Palis of the IMU, Jiang Zemin of China, and Shiing-Shen Chern of the 2002 congress (honorary). (Courtesy of the ICM 2002.)

Vladimir Voevodsky from the Institute for Advanced Study, for having "developed new cohomology theories for algebraic varieties, thereby providing new insights into number theory and algebraic geometry."

The Nevanlinna Prize Committee consisted of Michael Rabin, as chairman, and Andrei Agrachev, Ingrid Daubechies, Wolfgang Hackbusch, and Alexander Schrijver. The prize was awarded to Madhu Sudan from the Massachusetts Institute of Technology, for his "contribution to probabilistically checkable proofs, to non-approximability of optimization problems, and to error-correcting codes."

The Fields Medals were presented to Lafforque and Voevodsky by China's president Jiang Zemin—who, being an engineer, had a deep appreciation for mathematics. The moment was replete with music from loudspeakers and flashes of the press. More modest was

the presentation of the Nevanlinna Prize to Sudan by the secretary of the International Mathematical Union, Phillip Griffiths.

The ceremony ended with the laudations of the works of the awardees: Gérard Laumon spoke on the work of Laurent Lafforgue, Christophe Soulé on the work of Vladimir Voevodsky, and Shafi Goldwasser on the work of Madhu Sudan.

Two new mathematical prizes were announced. One was the Gauss Prize for applications of mathematics, to be jointly awarded every four years by the International Mathematical Union and the Deutsche Mathematiker-Vereinigung. The first award would be presented at the 2006 international congress. The other was the Abel Prize, created by the Norwegian Government on the occasion of the bicentenary of Niels Henrik Abel, to be awarded annually by the Norwegian

The recipients of the 2002 awards. (Courtesy of the ICM 2002.)

Academy of Science and Letters. This prize was meant to recognize contributions of extraordinary depth and influence to the mathematical sciences. The first award was scheduled for 2003.

After the opening ceremony, there was a multicourse banquet for 5000 people in the Banquet Hall of the Great Hall of the People .

Banquet of the 2002 congress. (From the ICM 2002.)

The scientific program took place in the gigantic Beijing International Conference Center. It consisted of 20 plenary lectures, 174 invited lectures, and more than 1200 short communications and poster sessions.

The 20 plenary lectures, in chronological order of delivery, were

- *"Chtoucas de Drinfeld, Formule des Traces d'Arthur-Selberg et Correspondance de Langlands,"* by Laurent Lafforgue;
- "Pattern Theory: The Mathematics of Perception," by David Mumford;
- "Geometry and Nonlinear Analysis," by Gang Tian;
- "Nonlinear Elliptic Theory and the Monge-Ampere Equation," by Luis Caffarelli;
- "Classification of Supersymmetries," by Victor Kac;
- "Random Matrices, Free Probability and the Invariant Subspace Problem Relative to a von Neumann Algebra," by Uffe Haagerup;
- "Discrete Mathematics: Methods and Challenges," by Noga Alon;
- "Mathematical Foundations of Modern Cryptography: Computational Complexity Perspective," by Shafi Goldwasser;

The lectures of the 2002 congress took place at the Beijing International Conference Center. (Courtesy the ICM 2002.)

- "Cohomology of Moduli Spaces," by Frances Kirwan;

- "Some Recent Transcendental Techniques in Algebraic and Complex Geometry," by Yum-Tong Siu;

- "Algebraic Topology and Modular Forms," by Michael J. Hopkins;

- "Differential Complexes and Numerical Stability," by Douglas N. Arnold;

- "Hyperbolic Systems of Conservation Laws in One Space Dimension," by Alberto Bressan;

- "Geometric Construction of Representations of Affine Algebras," by Hiraku Nakajima;

- "Galois Representations," by Richard Taylor;

- "Some Highlights of Percolation," by Harry Kesten;

- "Non-linear Partial Differential Equations in Conformal Geometry," by Sun-Yung Alice Chang and Paul C. Yang;

- "Emerging Applications of Geometric Multiscale Analysis," by David L. Donoho;

- "Knotted Solitons," by Ludwig Faddeev;

- "Singularities in String Theory," by Edward Witten.

Fields medalist Lafforgue in his plenary lecture raised an old theme of the international congresses: the language issue. He began by saying that he would prefer to speak French and complained that the "domination of the whole world by a single country —whatever its merits—by a single culture and by a single language can be very destructive of diversity of thought." He chose to address the congress with the following combination of three languages: the lecture was in English, and the transparencies were both in French and Chinese (any other combination would have produced a similar effect).

The invited lectures were distributed into the 19 classical scientific sections of the congresses, which were (in parentheses is the number of invited lectures in each section):

- Section 1: Logic (5),

- Section 2: Algebra (9),

- Section 3: Number Theory (10),
- Section 4: Differential Geometry (14),
- Section 5: Topology (9),
- Section 6: Algebraic and Complex Geometry (8),
- Section 7: Lie Groups and Representation Theory (11),
- Section 8: Real and Complex Analysis (8),
- Section 9: Operator Algebras and Functional Analysis (6),
- Section 10: Probability and Statistics (13),
- Section 11: Partial Differential Equations (13),
- Section 12: Ordinary Differential Equations and Dynamical Systems (11),
- Section 13: Mathematical Physics (12),
- Section 14: Combinatorics (8),
- Section 15: Mathematical Aspects of Computer Science (7),
- Section 16: Numerical Analysis and Scientific Computing (7),
- Section 17: Applications of Mathematics in the Sciences (11),
- Section 18: Mathematics Education and Popularization of Mathematics (5),
- Section 19: History of Mathematics (3).

There was a very large number of satellite conferences organized in connection with the international congress, 46. They were distributed all over China, in 26 different cities, as well as in six cities in Japan, Russia, Singapore, South Korea, and Vietnam.

As to other activities set up by the organizing committee, the chairman was proud to report that "three public lectures have attracted a broad social audience and were of great significance to the popularization of mathematics and its applications." He was referring to

three lectures by John F. Nash, Jr., Mary Poovey, and Wen-Tsun Wu aimed at the general public.

S.-S. Chern and John Nash, Jr., Nobel Prize in Economics, at the 2002 congress. (Courtesy of the ICM 2002.)

Nash, 1994 Nobel laureate in economics, lectured on Wednesday, August 21, at the congress site. His talk was entitled "Studying Cooperation in Games via Agencies." His presence was a a mass media event. Poovey, director of the Institute for the History of the Production of Knowledge of New York University, gave a lecture entitled "Can Numbers Ensure Honesty? Unrealistic Expectations and the U.S. Accounting Scandal." Wu's lecture was scheduled within an international symposium on the history of Chinese mathematics and was delivered at the China Science and Technology Museum. It was entitled "On the Development of the Real Number System in Ancient China." The symposium and the museum also hosted an exhibit of ancient Chinese mathematical toys.

Public attention to the congress profited from a public lecture by Stephen Hawking entitled "Brave New World." Hawking was in China for a satellite conference of the ICM. His lecture was followed by more than 2200 people.

Other singular activities were organized by the congress: a Juvenile Mathematics Forum, where 300 school boys from all over China attended the opening of the ICM and were engaged in other activities; and the ICM 2002 Mathematics Summer Campus, designed to raise the enthusiasm of the younger generation for mathematics.

We have already mentioned the impressive participation—4270 participants from 101 countries. The official figures for the largest national groups were: 1973 from China, 459 from the U.S.A., 191 from Japan, 167 from Russia, 107 from Germany, 95 from Korea (sic), 94 from France, and 90 from Canada. The distribution of participants by continents was 1 percent from Australia, 3 percent from Africa, 56 percent from Asia, 16 percent from the Americas, and 24 percent from Europe. Participants from developing countries represented 52 percent of the total number of participants. This huge congress required a huge financing and effort. The Chinese government gave 10,000,000 yuan, and the donations from universities, industries, and individuals were 3,000,000 yuan. In the local organizing committee, there were mathematicians from mainland China as well as representatives from Taiwan, Hong Kong, and Chinese overseas mathematicians. There were 300 student volunteers, mainly undergraduate mathematics students from Peking University.

The closing ceremony was held at the Beijing International Conference Center, on August 28. The congress was informed of decisions taken at the 14th General Assembly of the IMU held in Shanghai. One was the election of John M. Ball as president of the International Mathematical Union for the period 2003–2006. Regarding new projects, Ball explained the plan to retro-digitize the entire mathematical literature. For the first time, a woman was elected as a member of the executive committee, Ragni Piene from Norway, and for the first time a Chinese mathematician was elected as a member of the executive committee, Ma Zhiming, who was chairman of the organizing committee for ICM 2002.

On behalf of the Spanish IMU Committee, which represented all the Spanish mathematical societies, Carles Casacuberta issued the invitation to hold the 2006 congress in Madrid. (India, Italy, and Spain had sought the nomination.)

Let us end with some beautiful words of Ludwig Faddeev, from the Russian Academy of Sciences in St. Petersburg, who declared himself to be "a veteran of the ICM" (he stated attending in 1962). At the closing of the Beijing congress, he said, "The main idea of the ICM is to confirm the unity and universality of mathematics."

MADRID 2006

The Fields Medal Committee was privileged to consider a number of remarkable young mathematicians. Although the choice was a difficult one, the Committee was unanimous in selecting four medalists whose wonderful work demonstrates the breadth and richness of the subject. I will announce the names of the winners in alphabetical order.

A Fields Medal is awarded to Andrei Okounkov, of the Department of Mathematics, Princeton University, for his contributions bridging probability, representation theory and algebraic geometry.

A Fields Medal is awarded to Grigory Perelman, of St. Petersburg, for his contributions to geometry and his revolutionary insights into the analytical and geometric structure of the Ricci flow. I regret that Dr. Perelman has declined to accept the medal.

A Fields Medal is awarded to Terence Tao, of the Department of Mathematics, University of California at

The King of Spain with the Fields medalists (except for Perelman) and the Nevanlinna Prize winner, ICM 2006. (Courtesy of the ICM 2006.)

Los Angeles, for his contributions to partial differential equations, combinatorics, harmonic analysis and additive number theory.

A Fields Medal is awarded to Wendelin Werner, of the Laboratoire de Mathématiques, Université Paris-Sud, for his contributions to the development of stochastic Loewner evolution, the geometry of two-dimensional Brownian motion, and conformal field theory.

THIS WAS THE OFFICIAL ANNOUNCEMENT of the award winners made by Sir John Ball, president of the International Mathematical Union and chair of the Fields Medal Committee, at the opening ceremony of the 2006 congress. The other members of the committee were Enrico Arbarello, Jeff Cheeger, Donald Dawson, Gerhard Huisken, Curtis T. McMullen, Aleksei N. Parshin, Tom Spencer, and Michèle Vergne.

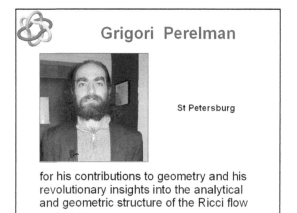

Grigori Perelman

St Petersburg

for his contributions to geometry and his revolutionary insights into the analytical and geometric structure of the Ricci flow

Grigory Perelman did not accept the 2006 Fields Medal. (Courtesy of the ICM 2006.)

Perelman's decision was unprecedented. For the first time in the history of the medal, an awardee had refused to accept the prize. Two months before, Ball had visited Perelman in St. Petersburg in a desperate attempt to persuade him to accept the medal, even suggesting that he need not attend the congress. But Perelman's decision was definite. He had already left his position at the Steklov Institute and had decided to withdraw from being a professional mathematician.

It is unquestionable that the "Perelman affair" caused the congress to receive extraordinary attention from the media. An hour before the opening ceremony, media reporters from all over the world were going around the congress venue asking everybody "Where is Perelman? Who is Perelman?" The attention paid to the congress reached the layman: taxi drivers, when asked to head for the Palacio Municipal de Congresos, the congress venue (see page 171), would enquire "Has Perelman finally showed up?" The debate on Perelman's attitude was heated with the publication in *The New Yorker* of a controversial article by Silvia Nasar (the author of *A Beautiful Mind*, on John Nash's life) and David Gruber on the circumstances surrounding Perelman's proof of Poincaré's Conjecture. All these events caused intense public awareness of the congress. However, for some mathematicians, there was the danger of once more spreading the image of the mathematician as an eccentric personality. The danger was real: despite the efforts of the congress press office (the Madrid congress was the first one to have a professional press office), the media preferred to focus on Perelman's disdain for awards. The media avoided speaking of his passion for opera or his devotion to taking longs walks though St. Petersburg or his avoidance of nasty academic quarrels.

Scientifically, the "Perelman affair" also dominated the beginning of the congress. As a substitute for Perelman's lecture, expectation was concentrated on the first plenary lecture of the congress, "The Poincaré Conjecture," by Richard Hamilton, who was the creator of the Ricci flow—the basic tool for the proof of the conjecture. The special lecture by John W. Morgan on "A Report on the Poincaré Conjecture" attracted much attention. It aimed to popularize the latest advances on

Press conference of the 2006 award winners (Wendelin Werner is missing). (Courtesy of the ICM 2006.)

the topic in a comprehensible way for a large scientific community (even for the educated layman!).

The opening ceremony continued with the awarding of the Nevanlinna Prize. The committee for the prize was chaired by Margaret H. Wright and consisted of Samson Abramsky, Franco Brezzi, Gert-Martin Greuel, and Johan Håstad. The award was granted to Jon M. Kleinberg from Cornell University, "for deep, creative and insightful contributions to the mathematical theory of the global information environment."

Applications of mathematics were a special feature of this congress. We see this in Kleinberg's work, which contained plenty of concrete practical applications, such as improving the effectiveness of web search engines or Internet routing. The special role in the congress of the applications of mathematics was emphasized by the awarding for the first time of the Carl

Friedrich Gauss Prize for Applications of Mathematics (we have already discussed the details of the award in "Awards of the ICM"). The prize was awarded to the preeminent Japanese mathematician Kiyoshi Itô for "laying the foundations of the theory of stochastic differential equations and stochastic analysis."

The committee for the prize (chaired by Martin Grötschel and consisting of Robert E. Bixby, Frank den Hollander, Stéphane Mallat, and Ian Sloan) considered that "Itô's work has emerged as one of the major mathematical innovations of the twentieth century and has found a wide range of applications outside of mathematics." These applications included areas such as engineering, physics, biology, economics, and finance.

The work of the award winners was reported to the congress in a series of twenty-minute laudations: Giovanni Felder spoke on Okounkov's work, John Lott

Left: Kiyoshi Itô's daughter receiving the 2006 Gauss Prize on behalf of her father. Right: The president of the IMU, Sir John Ball, presenting the Gauss Medal to Kiyoshi Itô, in Kyoto after ICM 2006. (Left: Courtesy of the ICM 2006; right: Courtesy of Junko Itô. Photo: Armin Mester.)

on Perelman's, Charles Fefferman on Tao's, Charles M. Newman on Werner's, and John Hopcroft on Kleinberg's. Hans Föllmer gave a forty-five minute lecture on "The Work of Kiyoshi Itô and Its Impact."

The opening ceremony was presided over by King Juan Carlos I of Spain, who also presided over the Honor Committee and presented the prizes to the award winners. For health reasons, Itô was unable to be present at the award ceremony, but his daughter Junko Itô, professor of linguistics at the University of California, accepted the prize on behalf of her father. Later, John Ball personally took the Gauss Medal to Kyoto and presented it to Kiyoshi Itô. The opening ceremony included a video showing the ancient relation of mathematics to art and culture in Spain, a musical performance by the Ara Malikian Trio, and a five-minute computer graphics video entitled *The Borromean Rings* about the new logo created by John Sullivan for the International Mathematical Union. After the ceremony, there was a reception, where the King of Spain socialized and chatted with the participants.

In the opening ceremony, there was a recognition of the community work that is performed by mathematicians. John Ball praised the effort of the Spanish mathematical community for the organization of the congress. He proposed Manuel de León as president of the congress and said:

> Mathematics has a strong record of service, freely given. We see this in the time and care spent in the refereeing of papers and other forms of peer review. We see it in the running of mathematical societies and journals, in the provision of free mathematical software and teaching resources, and in the various projects world-wide to improve electronic access to the mathematical literature, old and new. We see it in the nurturing of students beyond the call of duty.

The scientific part of the congress consisted of 20 plenary lectures, 169 invited talks distributed in 20 sections, and some 1000 short communications and posters (the organizing committee made a serious effort to encourage the presentation of posters, arranging a competition with prizes for the best posters). The list of scientific sections was similar to that of the Beijing 2002 congress except for the addition of a section on control theory and optimization, which had already existed at the Berlin 1998 congress.

The 20 plenary lectures were

- "Universality for Mathematical and Physical Systems," by Percy Deift;
- "Kähler Manifolds and Transcendental Techniques in Algebraic Geometry," by Jean-Pierre Demailly;
- "Optimal Computation," by Ronald A. DeVore;
- "Symplectic Field Theory and Its Applications," by Yakov Eliashberg;
- "Knots and Dynamics," by Étienne Ghys;
- "The Poincaré Conjecture," by Richard Hamilton;
- "Prime Numbers and L-functions," by Henryk Iwaniec;
- "High Dimensional Statistical Inference and Random Matrices," by Iain M. Johnstone;
- "Iwasawa Theory and Generalizations," by Kazuya Kato;
- "Energy-Driven Pattern Formation," by Robert V. Kohn;
- "Moduli Spaces from a Topological Viewpoint," by Ib Madsen;
- "Advances in Convex Optimization: Conic Programming," by Arkadi Nemirovski;
- "Deformation and Rigidity for Group Actions and von Neumann Algebras," by Sorin Popa;
- "Cardiovascular Mathematics," by Alfio Quarteroni;
- "Conformally Invariant Scaling Limits: An Overview and a Collection of Problems," by Oded Schramm;
- "Increasing and Decreasing Subsequences and Their Variants," by Richard P. Stanley;
- "The Dichotomy between Structure and Randomness, Arithmetic Progressions, and the Primes," by Terence Tao;
- "Perspectives in Nonlinear Diffusion: Between Analysis, Physics, and Geometry," by Juan Luis Vázquez;
- "Applications of Equivariant Cohomology," by Michèle Vergne;
- "P, NP, and Mathematics: A Computational Complexity Perspective," by Avi Wigderson.

One of these lectures, Hamilton's, is not included in the proceedings of the congress since it was not sent to the editors. There was also a series of 11 lectures on "Ibero-American Mathematics in the 19th and 20th Centuries" organized by the International Commission on the History of Mathematics.

There was a record number of 64 satellite conferences associated with the international congress. It is also remarkable that 28 of these took place outside Spain. Specifically, ten were held in Portugal, five in the U.K., and others in eight different European countries.

We have already seen that applications of mathematics were a special feature of the congress. The round-table discussion "Are Pure and Applied Mathematics Drifting Apart?" was organized on this topic. The discussion took as a starting point the fact that mathematics is broadening its scope, a process that causes the cross-fertilization between different fields of mathematics. The panelists for this interesting discussion (which can be read in the proceedings) were Lennart Carleson, Ronald Coifman, Yuri Manin, Helmut Neunzert, and Peter Sarnak; the moderator was the president of the International Mathematical Union, John Ball.

There were other activities organized at the congress or in connection with it. In some way or another, they relate to an interesting and deep statement made by John Ball at the opening of the congress: "Mathematicians do not own mathematics."

Lecture at 2006 congress. (Courtesy of the ICM 2006.)

One of these special activities was a panel discussion on "Should Mathematicians Care about Communicating to Broad Audiences?: Theory and Practice" organized by the European Mathematical Society. The discussion aimed at inviting mathematical researchers to reflect on their role in the battle for communicating about science, with the view that, in most countries, mathematics is not at present on a par in the media with other basic sciences. The panel was chaired by Jean-Pierre Bourguignon with the participation of Björn Engquist, Marcus du Sautoy, Alexei B. Sossinsky, François Tisseyre, and Philippe Tondeur.

Another activity was the school "Mathematics for Peace and Development" for young mathematicians from Arab countries (including Palestine), Latin America, Europe, and Israel. It was held at the University of Córdoba prior to the ICM. The basis for activity was the situation of Spain in history as a bridge between Africa, the Near East, Europe, and Latin America. The city of Córdoba was chosen as the venue since it was a symbol of the "Three Cultures," a place where Christians, Jews, and Muslims had lived side by side in tolerance for centuries.

An important element of the 2006 international congress was the number of activities on cultural dissemination. They were aimed at society in general, with the intention of highlighting the role played by mathematics in the culture of humanity throughout history. The most ambitious was the exhibition *The Life of Numbers* in the Spanish National Library, which we have already discussed in "Social Life at the ICM." Other exhibitions were: *Demoscene: Mathematics in Movement* in the Conde Duque Centro Cultural, showing a selection of computer-animated films that included live commentary by some of their creators; *History of Mathematical Knowledge* of mathematical texts since the sixteenth century, at the Historical Library of the Universidad Complutense de Madrid;

Benoit Mandelbrot chaired the jury that selected candidates for an international competition for the exhibition *Fractal Art: Beauty and Mathematics*, ICM 2006. (Left: Courtesy of the ICM 2006; right: *And how is your husband Mrs. Escher*, © Nada Kringels.)

and *Fractal Art: Beauty and Mathematics*, which displayed at the congress venue the work of the 28 finalists of an international contest organized by ICM 2006. Associated with the contest was the lecture "The Nature of Roughness in Mathematics, Science, and Art" by Benoit Mandelbrot, who headed the jury.

A spectacular event offered to participants was the live sculpting of a square block of black granite weighing several tons. The well-known Japanese sculptor Keizo Ushio worked facing the public on the sidewalk in front of the congress site. He first fashioned a torus, and then, with intense expectation by the congress par-

The Japanese sculptor Keizo Ushio created, in public at the 2006 congress site, a black granite sculpture featuring two intertwining halves of a split torus. (Courtesy of the ICM 2006.)

ticipants, he drilled the diameter of the meridian circle making a 360-degree rotation around the torus obtaining two intertwined halves, each doubly twisted. The final sculpture resembled a Baroque version of the symbol of infinity.

The congress continued with two of the long-standing traditions in the history of the ICM. One was the edition of a commemorative stamp, which we have seen in "Images of the ICM."

Another ICM tradition has been the interest in classical works in mathematics. This time, there was an edition of the works of Archimedes: *On the Sphere and the Cylinder*, *On the Measurement of the Circle*, and *The Quadrature of the Parabola*. The edition was a facsimile of a sixteenth-century manuscript, from the library of the Monastery of El Escorial, copied in Venice at the expense of Charles V's ambassador. The original manuscript is in the Marciana Library. Antonio J. Durán was in charge of the edition, published jointly with the Real Sociedad Matemática Española. It included a second volume with an annotated Spanish translation. The two-volume set was the official gift for the plenary lecturers and the invited speakers.

A special commemorative edition of the works of Archimedes was published on the occasion of the 2006 congress. It included a facsimile of a sixteenth-century manuscript from the Monastery of El Escorial Library containing the works *On the Sphere and Cylinder, Measurement of the Circle,* and *Quadrature of the Parabola.* (Courtesy of the ICM 2006 and the Real Sociedad Matemática Española.)

The organization of the congress was a joint venture of the main Spanish mathematical societies the Real Sociedad Matemática Española, the Societat Catalana de Matemàtiques, la Sociedad Española de Matemática Aplicada, and the Sociedad de Estadística e Investigación Operativa. Participation was 3425, of whom 1330 were from Spain (almost 40 percent); there were also 400 guests. The number of countries with participants set an all-time record of 108 (a figure from which the Falkland Islands, Hong Kong, and Netherlands Antilles should be deducted).

Participants by country*		
Algeria 11	Hungary 21	Panama 1
Argentina 32	India 111	Paraguay 1
Armenia 1	Indonesia 2	Peru 5
Australia 23	Iran 38	Phillippines 6
Austria 11	Ireland 8	Poland 32
Azerbaijan 11	Israel 34	Portugal 32
Bangladesh 1	Italy 81	P.R. China 33
Belarus 4	Ivory Coast 3	Puerto Rico 5
Belgium 12	Japan 128	Romania 22
Benin 1	Jordan 1	Russian Federation ... 75
Bolivia 1	Kazakhstan 4	Saudi Arabia 6
Bosnia & Herzegovina .. 3	Kenya 2	Senegal 2
Brazil 47	Kuwait 1	Serbia & Montenegro .. 15
Bulgaria 2	Kyrgyzstan 1	Singapore 6
Burkina Faso 1	Latvia 3	South Africa 12
Cambodia 1	Lesotho 2	South Corea 61
Cameroon 4	Lithuania 1	Spain 1330
Canada 53	Luxembourg 1	Sudan 1
Chile 9	Madagascar 1	Sweden 21
Colombia 13	Malaysia 3	Switzerland 26
Croatia 6	Mauritania 1	Tadjikistan 1
Cuba 2	Mexico 57	Taiwan 24
Cyprus 1	Moldova 2	Tanzania 1
Czech Republic 7	Mongolia 2	Thailand 4
Denmark 9	Morocco 15	Tunisia 7
Ecuador 1	Mozambique 2	Turkey 16
Egypt 6	Nederlandse Antillen .. 1	Turkish Cyprus 2
Estonia 6	Nepal 1	Uganda 1
Falkland Islands 1	Netherlands 25	Ukraine 8
Finland 26	New Zealand 3	United Arab Emirates ... 4
France 122	Nicaragua 1	United Kingdom 103
Georgia 4	Nigeria 4	USA 383
Germany 120	North Corea 3	Uruguay 8
Greece 15	Norway 16	Uzbekistan 1
Guatemala 1	Pakistan 4	Venezuela 11
Hong-Kong 1	Palestine 1	Vietnam 6

Participants by country at the 2006 congress. (From the proceedings of the 2006 ICM, European Mathematical Society 2006.)

At the closing ceremony, the congress was informed of the decisions taken at the General Assembly of the International Mathematical Union held in Santiago de Compostela. One was the election of László Lovász as

president of the union for the period 2007–2010. Another decision concerned the Digital Mathematical Library project, whose ultimate goal is to create a worldwide network of digital mathematical literature. (The congress included a new feature regarding the "digital era." The proceedings of the Beijing 2002 congress had included a DVD with half an hour of information. For this congress, the proceedings also included a DVD, which showed the opening and closing ceremonies, all plenary lectures, and the panel discussions.)

László Lovász, in his address to the congress, returned to a recurring theme of the international congresses:

> When one arrives at a Congress, one cannot feel but overwhelmed by the number of people and by the variety of mathematics that is presented here. One could walk the corridors for minutes without seeing a familiar face, and one could browse the abstracts long before seeing a topic that one, say, did research in. This is so even for a senior person who attended many previous Congresses, and obviously a young person who has not been to previous Congresses must feel this even more.
>
> It is perhaps because of this feeling that people repeatedly bring up the idea of abandoning these International Congresses. I feel this would be a serious mistake. I talked to scientists working in other fields, and they expressed their envy for the fact that we have a meeting where the best mathematicians tell to all of us what are the main problems, trends, or paradigms of their fields; where we honor the recipients of major prizes, and hear and discuss their work; where we have panel discussions and also corridor discussions about important issues facing our science or our community.

The congress ended with the invitation by the Indian representative Rajat Tandon to hold the 2010 International congress in the city of Hyderabad:

> Hyderabad, like Madrid, is a wonderful composition of the old and the new. This city, founded more than 400 years ago, houses teeming bazaars, old jewelry and

The next ICM will be held in India: Hyderabad 2010, August 19–27. (Courtesy of the ICM 2006.)

The exhibition *The ICM through History*. (Courtesy of the ICM 2006.)

fine craftsmen, old forts and mausoleums. Cosmopolitan in its population you find people of all faiths living and learning together here.

...

Two hundred years ago this city expanded to the twin cities of Hyderabad and Secunderabad with the addition of a cantonment area and today greater Hyderabad is a conglomeration of three cities in one with the modern Cyberabad area which is second only to Bangalore as the information technology heart of India.

We end this account of the ICMs as we started: with the exhibition *The ICM through History*. The organizing committee decided to commemorate the 25 ICMs held from the first in Zurich in 1897 to the Madrid congress in 2006 with an exhibition. The aim of the exhibition was to provide a visual chronicle of all the ICMs, emphasizing their significance in terms of human endeavor and using the activities of mathematicians at the ICMs as a mirror in which his-

tory, culture, technology, fashion, and changing attitudes were reflected. Some 500 written and photographic documents provided a twin portrait of the ICMs; on the one hand, there was a chronological review of the history of the ICM, and on the other a transversal view through the social life of the congresses, the graphic design for the congresses, and the buildings where they were held. The physical and conceptual heart of the exhibition resided in the display of the medals, original reproductions of the Fields, Nevanlinna, and Gauss awards, provided by the Royal Canadian Mint, the University of Helsinki, and the Deutsche Mathematiker-Vereinigung, respectively. Materials were provided by many universities, libraries, archives, museums, mathematical societies, and individuals, which made it possible to assemble an extraordinary collection of photographs and documents, many of these never available to the public before.

Institut de France
Académie des Sciences

Paris, le 8 Juin 1921,

Mon cher collègue,

J'ai reçu votre envoi des statuts de l'Union Mathématique. Le Comité exécutif du Conseil international de recherches, qui s'est réuni hier, les a approuvés.

J'ai eu de bonnes nouvelles de vous par M. Lecointe, actuellement à Paris.

Veuillez croire à mon bien cordial souvenir,

Em Picard

J'ai transmis les statuts à M. Schuster.

Communication of the approval of the statutes of the old Union Mathématique Internationale by the International Research Council. (Courtesy of Olli Lehto.)

Coda

INTERNATIONAL MATHEMATICAL UNION

THE INTERNATIONAL MATHEMATICAL Union is an offspring of the International Congress of Mathematicians. This is as it should be. The reverse situation would have had all the weaknesses of a purely administrative decision. However, as we have seen, the development of the union was complicated by the aftereffects of World War I. This resulted in a precarious start in life for the union and almost caused the death of the congresses. Fortunately, the deep commitment to international collaboration among mathematicians allowed for a successful rebirth of the union after World War II. Later, the union grew healthy and strong and provided strong leadership for the congresses, guaranteeing a sound future for both. The history of the union is explained in detail in Olli Lehto's book *Mathematics without Borders*. We end with a chronology of the landmarks in the history of the union.

1908 The International Congress in Rome agrees to propose the establishment of an International Association of Mathematicians to the next Congress.

1912 The International Congress in Cambridge decides that "our existing arrangements for periodical Congresses meet the requirements of the case."

1919 The Allied Powers create in Brussels the International Research Council (IRC) with the mandate of promoting international scientific associations, excluding the Central Powers.

1920 The Union Mathématique Internationale (UMI) is created at the Strasbourg congress. The founding countries are: Belgium, Czechoslovakia, France, Greece, Italy, Japan, Poland, Portugal, Serbia, the United Kingdom, and the United States of America. Countries not members of the IRC are not allowed to join.

1924 The Toronto congress is held under IRC regulations: German, Austrian, Hungarian, and Bulgarian mathematicians are not invited.

1928 Breaking the union's opposition, the Bologna congress is open to all mathematicians.

1931 The union's statutes expire. After 12 years of controversial existence, the IRC is superseded by the International Council of Scientific Unions (ICSU).

1932 The union is dissolved. At the Zurich congress, it is decided that "an international commission is formed in order to re-study the question of the international collaborations in the sphere of mathematics."

1936 By decision of the Oslo congress, the issue of international collaboration is abandoned.

1950 A Constitutive Convention in New York decides to create the International Mathematical Union (IMU) based on the principle of unrestricted internationalism.

1952 Union's first General Assembly in Rome with 22 countries: Argentina, Australia, Austria, Belgium, Canada, Cuba, Denmark, Finland, France, Germany, Greece, Italy, Japan, the Netherlands, Norway, Pakistan, Peru, Spain, Switzerland, the United Kingdom, the United States of America, and Yugoslavia.

1954 The old Commission Internationale de L'Enseignement Mathématique, which was created in the 1908 Congress with a four-year mandate, renewed in 1912, 1928, 1932, and 1936, becomes part of the union and is renamed the International Commission on Mathematical Instruction (ICMI).

1957 The Soviet Union and other socialist European countries join the International Mathematical Union.

1958 Realizing an idea of the Zurich 1897 congress, the first *World Directory of Mathematicians* is published by the union and the Tata Institute. It contains 3500 names of active mathematicians. It is renewed every four years until 2002.

1962 IMU-ICM marriage: The union takes under its control the scientific program of the congresses and the awarding of prizes.

1969 First ICME of ICMI: The first International Congress on Mathematical Education takes place in Lyon. They continue to be held every four years, between the ICMs.

1979 Opening to the world: The International Mathematical Union creates the Commission on Development and Exchange to promote mathematics in developing countries.

1986 The People's Republic of China joins the union.

1987 Looking to the past: the International Commission on the History of Mathematics is created, jointly by the International Mathematical Union and the International Union of History and Philosophy of Science.

2000 Celebration of the World Mathematical Year 2000, proposed by the International Mathematical Union in 1992.

ACKNOWLEDGMENTS

INSTITUTIONS

AIP Emilio Segre Visual Archives; American Mathematical Society; Archiv der Universität Wien; Archives Henri Poincaré, Université Nancy 2, CNRS; Archivio Storico dell' Accademia Nazionale delle Scienze; Archivio Storico e Archivio Fotografico, Universitá di Bologna; Biblioteca Comunale dell' Archiginnasio di Bologna; Biblioteca de CC. Matemáticas, Universidad Complutense de Madrid; Dipartimento di Matematica, Universitá di Bologna; Istituto Guido Castelnuovo, Universitá "La Sapienza" di Roma; Centrum för Vetenskapshistoria, Stockholm; Centrum voor Wiskunde en Informatica, Amsterdam; Department of Mathematics and Statistics, University of Helsinki; Department of Mathematics, University of British Columbia; Department of Mathematics, University of Toronto; Deutsche Mathematiker-Vereinigung; Eidgenössische Technische Hochschule-Bibliothek; Faculty of Mathematics, Ivanovo State University; Fields Institute for Research in Mathematical Sciences; Harvard University Archives; Institut Mittag-Leffler; Institute for Advanced Study, Princeton; Integral de Arte y Exposiciones, Granada; Kyoto International Conference Hall; Museo di Fisica, Dipartimento di Fisica, Universitá "La Sapienza" di Roma; Niedersaechsische Staats- und Universitätsbibliothek Göttingen; Royal Canadian Mint; Center for History of Science of the Royal Swedish Academy of Sciences; Ryhiner Collection, Stadt- und Universitätsbibliothek Bern; Science and Society Picture Library; Universitätsarchiv Heidelberg; Universitätsbibliothek Heidelberg; University of British Columbia Archives; and University of Helsinki Archives.

INDIVIDUALS

Don Albers, Gerald L. Alexanderson, Bogdan Bojarski, Benedetto Bongiorno, N. G. de Bruijn, J. J. Burckhardt, Vladimir Buyarov, Lennart Carleson, Mirta Castro Smirnova, Afton H. Cayford, Mary Adah Costello, Guillermo Curbera Suárez, Ana Curbera, Jaime Curbera, Anna Doubova, José Ferreirós, Linda Geraci, Livia Maria Giacardi, Natividad Gómez Pérez, Martin Grötschel, Hans Jarchow, Yafen Jin, Nils Voje Johansen, Jean-Pierre Kahane, Serge I. Khashin, Annette Kik, Harro Klein, Javier Lerín, Geza Makay, Gregori Margulis, N. N. Molchanov, Susumu Okada, Czeslaw Olech, Ben de Pagter, Constance Reid, Ricker-Kollmann family, Kristian Seip, Arild Stubhaug, Herman J. J. te Riele, Javier Soria, Jaap Top, Kenji Ueno, Juan Luis Varona, Juan Carlos Vega Lozano, Serena Vergano, Lucilla Vespucci, and Jaroslav Zemanek.

BIBLIOGRAPHY

CONGRESS PROCEEDINGS

Mathematical Papers Read at the International Mathematical Congress held in Connection with the World's Columbian Exposition Chicago 1893, Moore, E.H., Bolza, O. and Maschke, H. (eds.) New York: American Mathematical Society, 1896.

Verhandlungen des ersten Internationalen Mathematiker-Kongresses: in Zürich vom 9. bis 11. August 1897, Rudio, F. (ed.) Leipzig: B.G. Teubner, 1898.

Compte rendu du Deuxième Congrès Internationale des Mathématiciens, Duporcq, E. (ed.) Paris: Gauthier-Villars, 1902.

Verhandlungen des dritten Internationalen Mathematiker-Kongresses in Heidelberg: vom 8. bis 13. August 1904, Krazer, A. (ed.) Leipzig: Teubner, 1905.

Atti del IV Congresso internazionale dei matematici: Roma 6–11 Aprile 1908, Castelnuovo, G. (ed.) Roma: R. Accademia dei Lincei, 1909.

Proceedings of the Fifth International Congress of Mathematicians, Hobson, E.W. and Love, A.E.H. (eds.) Cambridge, UK: Cambridge University Press, 1913.

Comptes rendus du congrés international des mathématiciens: Strasbourg, 22–30 Sept., 1920, Villat, H. (ed.) Toulouse: Eduard Privat, 1921.

Proceedings of the International Mathematical Congress held in Toronto, August 11–16, 1924, Fields, J.C. (ed.) Toronto: The University of Toronto Press, 1928.

Atti del Congresso Internazionale dei Matematici: Bologna, 3–10 Settembre 1928 (VI). Bologna: Nicola Zanichelli, 1929–1932.

Verhandlungen des Internationalen Mathematiker-Kongresses Zürich 1932, Saxer, W. (ed.) Zurich: Orell Füssli, 1932.

Comptes rendus du congrés international des mathématiciens, Oslo 1936. Oslo: A.W. Brogers Boktrykkeri A/S, 1937.

Proceedings of the International Congress of Mathematicians, Cambridge, Massachusetts, 1950, Providence, RI: American Mathematical Society, 1952.

Proceedings of the International Congress of Mathematicians: 1954, Amsterdam, September 2–9, Gerretsen, J.C.H. and De Groot, J. (eds.) Amsterdam: North-Holland, 1957.

Proceedings of the International Congress of Mathematicians: 14–21 August 1958, Todd, J.A. (ed.) Cambridge, UK: Cambridge University Press, 1960.

Proceedings of the International Congress of Mathematicians: 15–22 August 1962. Djursholm: Institut Mittag-Leffler, 1962.

Truduy Mezdynarodnogo Kongressa Matematikov: Mockba—1966. Moscow: Mir, 1986.

Actes du Congrès International des Mathématiciens 1970. Paris: Gauthier-Villars, 1971.

Proceedings of the International Congress of Mathematicians, Vancouver, 1974, August 21–29, James, R.D. (ed.) Vancouver: Canadian Mathematical Congress, 1975.

Proceedings of the International Congress of Mathematicians: Helsinki 1978, Lehto, O. (ed.) Helsinki: Academia Scientiarum Fennica, 1980.

Proceedings of the International Congress of Mathematicians: August 16–24 1983: Warszawa, Ciesielski, Z. and Olech, C. (eds.) Warsaw: PWN-Polish Scientific, 1984.

Proceedings of the International Congress of Mathematicians, 1986: August 3–11, 1986, Berkeley, California, Gleason, A. M. (ed.) Providence, RI: American Mathematical Society, 1987.

Proceedings of the International Congress of Mathematicians: August 21–29, 1990, Kyoto, Japan, Satake, I. (ed.) Tokyo: Springer-Verlag, 1991.

Proceedings of the International Congress of Mathematicians: Zürich 1994, Chatterji, S.D. (ed.) Basel: Birkhäuser Verlag, 1995.

Proceedings of the International Congress of Mathematicians: Berlin 1998, August 18–27. Bielefeld: Documenta Mathematica, 1998.

Proceedings of the International Congress of Mathematicians: Beijing 2002, August 20–28, Li Tatsien , (ed.) Beijing: Higher Education Press, 2002.

Proceedings oh the International Congress of Mathematicians: Madrid, August 22-30, 2006, Sanz-Solé, M., Soria, J., Varona, J.L., and Verdera, J. (eds.) Zurich: European Mathematical Society, 2006.

Goldstein, C., Gray, J. and Ritter, J. *L'Europe Mathématique: Histories, Mythes, Identités*. Paris: Éditions de la Maison des Sciences de L'homme, 1996.

Goldstein, C., Schappacher, N. and Schwermer, J. *The Shaping of Arithmetic: After C. F. Gauss's Disquisitiones Arithmeticae*. New York: Springer-Verlag, 2007.

Greenaway, F. *Science International: A History of the International Research Council*. Cambridge, UK: Cambridge University Press, 1996.

Lehto, O. *Mathematics without Borders: A History of the International Mathematical Union*. New York: Springer-Verlag, 1988.

Monastyrsky, M. *Modern Mathematics in the Light of the Fields Medals*. Wellesley, MA: A K Peters, 1997.

Parshall, K.U. and Rice, A.C. (eds.) *Mathematics Unbound: The Evolution of an International Mathematical Research Community, 1800–1945*. Providence, RI: American Mathematical Society, 2002.

Reid, C. *Hilbert*. Berlin: Springer-Verlag, 1970.

Trostnikov, V.N. *Vsemirnii Kongress Matematikov v Moskve*. Moscow: Znanie, 1967.

BOOKS

Albers, D.J., Alexanderson, G.L., and Reid, C. *International Mathematical Congresses: An Illustrated History 1893–1986*. New York: Springer-Verlag, 1987.

Alexanderson, G.L. (ed.) *The Pólya Picture Album: Encounters of a Mathematician*. Boston: Birkhäuser, 1987.

Archibald, R.C. *A Semicentennial History of the American Mathematical Society*. New York: American Mathematical Society, 1938.

Cattermole, M.J.G. *Horace Darwin's Shop: A History of the Cambridge Scientific Instrument Company*. Bristol: Adam Hilger, 1987.

Dauben, J.W. *Georg Cantor: His Mathematics and Philosophy of the Infinite*. Cambridge, MA: Harvard University Press, 1979.

ARTICLES

Barnes, M.E. "The Man behind the Medal: A Glimpse at John Charles Fields." *The Australian Mathematical Society Gazette*, vol. 30, no. 5 (2003), 278–283.

Barrow-Green, J. "International Congresses of Mathematicians from Zurich 1897 to Cambridge 1912." *The Mathematical Intelligencer*, vol. 16, no. 2 (1994), 38–41.

Boltyanskii, V.G., Karmanov, V.G. and Rozov N.Kh. "International Congress of Mathematicians (Moscow, 1966)." *Uspekhi Matematicheskikh Nauk*, vol. 22, no. 4 (136) (1967), 137–143.

Cartwright, M. "Note on A.G. Cock's Paper 'Chauvinism in Science': The International Research Council, 1919–1926." *Notes and Records of the Royal Society of London*, vol. 39, no. 1 (1984), 125–128.

Crathorne, A.R. "The Fifth International Congress of Mathematicians." *Science N.S.*, vol. 36, no. 932 (1912), 622–624.

Cole, F.N. "Mathematical Papers Read at the International Mathematical Congress held in Connection with the World's Columbian Exhibition, Chicago, 1893." *Science N.S.*, vol. 4, no. 85 (1896), 200–201.

Dauben, J.W. "Mathematicians and World War I: The International Diplomacy of G. H. Hardy and Gösta Mittag-Leffler as Reflected in Their Personal Correspondence." *Historia Mathematica*, vol. 7, no. 3 (1980), 261–288.

Dresden, A. "The International Congress at Toronto." *Bulletin of the American Mathematical Society*, vol. 31 (1922), 1–10.

Dunnington, G.W. "Gauss, His *Disquisitiones Arithmeticae*, and His Comtemporaries in the Institut de France." *National Mathematics Magazine*, vol. 9, no. 7 (1935), 187–192.

Dunnington, G.W. "Oslo under the Integral Sign." *National Mathematics Magazine*, vol. 11, no. 2 (1936), 85–94.

Fehr, H. "Congrès international des mathematiciens, Paris, août 1900." *L'Enseignement mathématique*, vol. 2, no. 1 (1901), 378–382.

Fehr, H. "5e Congrès international des mathematiciens." *L'Enseignement mathématique*, vol. 14 (1912), 301–306.

Fehr, H. "Congrès international des mathematiciens." *L'Enseignement mathématique*, vol. 21 (1920), 192–207.

Fehr, H. "Commission internationale de l'Enseignement mathématique." *L'Enseignement mathématique*, vol. 21 (1920), 137–138.

Fehr, H. "Le congrès de Bologne." *L'Enseignement mathématique*, vol. 28 (1928), 28–53.

Fehr, H. "M. J.C. Fields." *L'Enseignement mathématique*, vol. 31 (1932), 127.

Fehr, H. "Union international des mathematiciens." *L'Enseignement mathématique*, vol. 31 (1932), 276–278.

Fehr, H. "Deux prix internationals de mathématiques." *L'Enseignement mathématique*, vol. 31 (1932), 279.

Fehr, H. "Le 10e congrès international des mathématiciens, Oslo 1936." *L'Enseignement mathématique*, vol. 35 (1936), 373–385.

Gårding, L. and Hörmander, L. "Why Is There No Nobel Prize in Mathematics?." *The Mathematical Intelligencer*, vol. 7, no. 3 (1985), 73–74.

Gowers, W.T. "The Two Cultures of Mathematics." In *Mathematics: Frontiers and Perspectives*, Arnold, V.I., Atiyah, M., Lax, P. and Mazur, B. (eds.), pp. 65–78. Providence, RI: American Mathematical Society, 1999.

Grattan-Guiness, I. "A Sideways Look at Hilbert's Twenty-three Problems of 1900." *Notices of the American Mathematical Society*, vol. 47, no. 7 (2000), 752–757.

Gray, J. "Who Would Have Won the Fields Medal One Hundred Years Ago?." *The Mathematical Intelligencer*, vol. 7, no. 3 (1985), 10–19.

Halsted, G.B. "The International Mathematical Congress." *Science N.S.*, vol. 1, no. 18 (1895), 486–487.

Halsted, G.B. "The International Mathematical Congress." *Science N.S.*, vol. 6, no. 141 (1897), 402–403.

Halsted, G.B. "The International Mathematical Congress." *The American Mathematical Monthly*, vol. 4, no. 8/9 (1897), 229–230.

Halsted, G.B. "The International Congress of Mathematicians." *The American Mathematical Monthly*, vol. 7, no. 8/9 (1897), 188-189.

Hardy, G.H. "International Congresses." *Science N.S.*, vol. 60, no. 1565 (1924), 591–592.

Hilbert, D. "Mathematical Problems." *Bulletin of the American Mathematical Society*, vol. 8 (1902), 437–479.

Howson, A.G. "Seventy-five Years of the International Commission on Mathematical Instruction." *Educational Studies in Mathematics*, vol. 15, no. 3 (1984), 75–93.

Kenschaft, P.C. "Charlotte Angas Scott, 1858–1931." *The College Mathematics Magazine*, vol. 18, no. 2 (1987), 98–110.

Lange, L.H. "I.C.M. Helsinki—Some Personal Notes." *The Two-Year College Mathematics Journal*, vol. 11, no. 2 (1980), 144–146.

Lorch, L. "Mathematics: International Congress." *Science*, vol. 155, no. 3765 (1967), 1038–1039.

Lorch, L. "Congress of Mathematicians in Moscow." *Uspekhi Matematicheskikh Nauk*, vol. 22, no. 4 (136) (1967), 144–146.

McLennan, J.C. "Prof. J.C. Fields, F.R.S." *Nature*, vol. 130 (1932), 688–689.

Monastyrsky, M. "Some Trends in Modern Mathematics and the Fields Medal." *Canadian Mathematical Society. CMS Notes*, vol. 33, no. 2, 1–5; no. 3 (2001), 6–9.

Moore, C.L.E. "The Fourth International Congress of Mathematicians." *Bulletin of the American Mathematical Society*, vol. 14 (1908), 481–498.

Morse, M. "The International Congress in Oslo." *Bulletin of the American Mathematical Society*, vol. 42 (1936), 777–781.

Mostow, G.D. "The 1983 Warsaw Congress of IMU." *Notices of the American Mathematical Society*, vol. 31, no. 10 (1983), 571–573.

Nasar, S. and Gruber, D. "Manifold Destiny: A Legendary Problem and a Battle over Who Solved It." *The New Yorker*, August 28 (2006), 44–57.

Parshall, K.H. "Mathematics in National Contexts (1875–1900): An International Overview. In *Proceedings of the International Congress of Mathematicians: Zürich 1994*, Chatterji, S.D. (ed.), pp. 1581–1591. Basel: Birkhäuser Verlag, 1995.

Parshall, K.H. and Rowe, D.E. "Embedded in the Culture: Mathematics at the World's Columbian Exposition of 1893." *The Mathematical Intelligencer*, vol. 15, no. 2 (1983), 40–45

Riehm, C. "The Early History of the Fields Medal." *Notices of the American Mathematical Society*, vol. 49, no. 7 (2002), 778–782.

Richardson, R.G.D. "International Congress of Mathematicians, Zurich 1932." *Bulletin of the American Mathematical Society*, vol. 38 (1932), 769–774.

Scott, C.A. "The International Congress of Mathematicians in Paris." *Bulletin of the American Mathematical Society*, vol. 7 (1901), 57–79.

Scott, C.A. "Compte Rendu du Deuxieme Congrès International des mathematiciens tenu a Paris by E. Duporcq." *Bulletin of the American Mathematical Society*, vol. 9 (1903), 214–215.

Smale, S. "On the Steps of Moscow University." *The Mathematical Intelligencer*, vol. 6, no. 2 (1984), 21–27.

Smith, D.E. "The Mathematical Congress at Strasbourg." *Bulletin of the American Mathematical Society*, vol. 26 (1920), 104–108.

Smith, D.E. "The International Congress of Mathematicians, Zurich, September 4–12, 1932." *Science N.S.*, vol. 76, no. 1977 (1932), 468–471.

Snyder, V. "The Fifth International Congress of Mathematicians." *Bulletin of the American Mathematical Society*, vol. 19 (1912), 107–130

Synge, J.L. "Obituary Notice of John Charles Fields." *Obituary Notices of Fellows of the Royal Society*, vol. 2 (1933), 130–138.

Synge, J.L. "John Charles Fields." *Journal of the London Mathematical Society*, vol. 8 (1933), 153–160.

Tropp, H.S. "The Origins and History of the Fields Medal." *Historia Mathematica*, vol. 3 (1976), 167–181.

Tyler, H.W. "The International Congress of Mathematicians at Heidelberg." *Bulletin of the American Mathematical Society*, vol. 11 (1905), 191–205.

Young, J.W.A. "The Fifth International Congress of Mathematicians." *The American Mathematical Monthly*, vol. 19, no. 10/11 (1912), 161–166.

NOTES AND ANNOUNCEMENTS IN JOURNALS

"The International Congress of Mathematicians." *Nature*, vol. 2, no. 1609 (1900), 418–420.

"International Congress of Mathematicians." *Science N.S.*, vol. 36, no. 923 (1912), 312.

"Unions scientifiques internationales." *L'Enseignement mathématique*, vol. 21 (1920), 59–60.

"Congrès international des mathematiciens." *L'Enseignement mathématique*, vol. 21 (1920), 60–61.

"Congrès international des mathematiciens." *L'Enseignement mathématique*, vol. 26 (1927), 321–322.

"Congrès international des mathematiciens." *L'Enseignement mathématique*, vol. 27 (1927), 146.

"The International Congress of Mathematicians." *Bulletin of the American Mathematical Society*, vol. 34 (1928), 533, 792.

"Congrès international des mathematiciens." *L'Enseignement mathématique*, vol. 31 (1932), 124–125.

"International Mathematical Congress Medals." *Science N.S.*, (1933) vol. 78, no. 2034, 573–574.

"Scientific Notes and News." *Science N.S.*, vol. 84, no. 2178 (1936), 287.

"International Mathematical Congress." *Science N.S.*, vol. 88, no. 2294 (1936), 571–572.

"International Congress of Mathematicians." *Bulletin of the American Mathematical Society*, vol. 44 (1938), 593–595

Bulletin of the American Mathematical Society, vol. 45, no. 11 (1939), 804–805.

"International Congress of Mathematicians." *Bulletin of the American Mathematical Society*, vol. 54 (1948), 599–601.

"International Congress of Mathematicians." *Bulletin of the American Mathematical Society*, vol. 56 (1950), 1–4.

"International Congress of Mathematicians." *Bulletin of the American Mathematical Society*, vol. 56 (1950), 217.

"International Congress of Mathematicians." *The American Mathematical Monthly*, vol. 57, no. 3 (1950), 147–150.

"International Congress of Mathematicians 1986." *Notices of the American Mathematical Society*, vol. 34, no. 10 (1983), 735–737.

INDEX

Abel, Niels Henrik, 67, 102, 103, 288
Abramsky, Samson, 295
academies and learned societies
    Académie de Paris, 1, 23, 51,
        109
    Academy of Finland, 118
    Accademia dei Lincei, 34, 41,
        127
    Berlin Academy, 1, 28, 29, 34,
        109, 150
    Cambridge Philosophical
        Society, 45, 47, 229
    Gesellschaft Deutscher
        Naturforscher und Ärzte, 5
    Irish Academy, 80
    Norwegian Academy, 288
    Polska Akademia Nauk, 207,
        216
    Royal Astronomical Society, 48,
        50
    Royal Society of London, 1, 51,
        109, 115, 116, 139
    Royal Swedish Academy, xii, 53
    Russian Academy, 77, 292
    Schweizerischen
        Naturforschenden
        Gesellschaft, 13, 51
    St. Petersburg Academy, 1, 34
    U.S. National Academy, 216,
        217, 224
    USSR Academy, 149, 175, 182,
        212
Adams, John F., 178, 194

Agrachev, Andrei, 288
Ahlfors, Lars Valerian, 101, 102, 104,
        117, 118, 128, 145, 201,
        219–221, 284
Aiken, Howard, 131
Albers, Donald J., xi, 223
Albert, Abraham Adrian, 130, 188
Alexander, James W., 93
Alexanderson, Gerald L., xi, 223
Alexandroff, Pavel Sergeyevich, 147
Alexandrov, Aleksandr Danilovic, 140
Alexandrov, Pavel Sergeyevich, 98,
        106, 136, 145, 150
Alon, Noga, 289
Amoroso, Luigi, 86
applied mathematics, 16, 32, 37, 38,
        40, 47–50, 77–79, 87, 95,
        104, 121–123, 130, 141,
        147, 255, 279, 287, 288,
        291, 295, 297
Araki, Huzihiro, 210, 211
Arbarello, Enrico, 294
Archimedes, 1, 2, 109–111, 119, 300
Arnold, Douglas N., 290
Arnold, Vladimir Igorevich, 196, 211
Arthur, James, 287
Artin, Emil, 145
Artin, Michael, 178
Atiyah, Michael Francis, 117, 177,
        178, 186, 221, 260, 261
authorities
    Archduke of Baden, 27, 32, 239,
        275

Benito Mussolini, 88, 106, 239
Emperor Hirohito, 265
General Wojciech Jaruzelski, 212
German Federal President
        Roman Herzog, 275, 282
Joseph Stalin, 150
Kaiser Wilhelm II, 35, 275
King Gustav VI Adolf of
        Sweden, xi, xii, 143, 145
King Haakon VII of Norway,
        101, 102
King Juan Carlos I of Spain,
        293, 296
King Vittorio Emanuele III of
        Italy, 37
Lieutenant Governor of British
        Columbia, 194, 195
President of China Jiang Zemin,
        285, 288
President of the French Republic
        Georges Pompidou, 190
Prince Philip, Duke of
        Edinburgh, 139
Queen Juliana of the
        Netherlands, 134
U.S. President Harry Truman,
        149
U.S. President Ronald Reagan,
        220, 221
awards
    Abel Prize, 110, 288
    Copley Medal, 109
    De Morgan Medal, 109

Grand Prix, 109
Medaglia Guccia, 41, 42, 44
negative effects, 110
Nobel Prize, 39, 48, 102, 110, 116, 291
Steiner Prize, 109

Bôcher, Maxime, 49
Babai, László, 270
Baker, Alan, 117, 185, 186
Ball, Sir John, xiii, 277, 292, 294, 296, 297
Banach, Stefan, 104, 205
banquets, balls, and parties, 15, 16, 25, 44, 54, 67, 72, 79, 89, 93, 131, 134, 142, 202, 216, 227, 228, 232, 234, 236, 238–246, 248, 266, 289
Bauer, Heinz, 196
Bellman, Richard, 178
Bendixson, Ivar, 15
Bernoulli, Daniel, 1, 13, 55
Bernoulli, Jacob, 13, 55
Bernoulli, Johann, 13, 55, 105, 273
Bernoulli, Nicolas, 1
Bernstein, Sergi, 93
Beurling, Arne, 128
Bianchi, Luigi, 87
Bieberbach, Ludwig, 84, 91–93
Birkhoff, George David, 89, 97, 101, 104, 125
Birman, Joan, 260
Bismut, Jean-Michel, 260, 279
Bixby, Robert E., 295
Bjerknes, Vilhelm, 104
Blaserna, Pietro, 41
Bloch, Spencer, 261, 287
Blumenthal, Otto, 282
Bochner, Salomon, 128
Bogolyubov, N. N, 140, 210
Bohr, Harald, 93, 98, 113, 128, 129
Bois-Reymond, Emil du, 20
Bollobás, Béla, 278
Bolza, Oskar, 7

Bombelli, Rafael, 87
Bombieri, Enrico, 117, 194–196, 260
Bompiani, E., 134
Bonaparte, Prince Roland, 23
books at ICM, 18, 32, 33, 51, 52, 80, 87, 94, 136, 141, 223, 253, 283, 300
Borcherds, Richard Ewen, 117, 277
Borel, Émile, 15, 49, 86
Borel, Armand, 145
Borsuk, Karol, 128, 136, 150
Bortolotti, Ettore, 87
Bott, Raoul, 186
Bourbaki, Nicolas, 176, 189, 195
Bourgain, Jean, 117, 269, 270, 287
Bourguignon, Jean-Pierre, 298
Branges, Louis de, 222
Brauer, Richard, 135, 186
Bressan, Alberto, 290
Brezzi, Franco, 295
Brioschi, Francesco, 5, 18, 23, 29
Browder, William, 186
Brown, Ernest W., 49
Bruijn, Nicolaas Govert de, 251, 252
Brunn, Hermann, 15
buildings, 155–172
    Beijing International Conference Center, 290
    Concertgebouw, 133, 134, 163
    Eidgenössiche Technische Hochschule, 9, 13, 56, 91, 92, 94, 155, 161, 169, 270
    Finlandia Hall, 165, 199
    Great Hall of the People, 171, 285, 286
    International Conference Center of Berlin, 170, 276
    Kyoto International Conference Hall, 168, 259
    Monastery of El Escorial, 62, 253, 300
    Palace of Culture, Warsaw, 167, 212
    Palacio Municipal de Congresos of Madrid, 171

Tonhalle, 12, 13, 92, 227, 245, 246
Zurich's Kongresshaus, 268–270
Burckhardt, J. J., xii, xiii, 94, 272, 273
Bureau, F., 134

Caffarelli, Luis, 269, 270, 289
calculating machines, mathematical instruments, computers, models, 32, 33, 51, 52, 94, 95, 119, 131, 136, 146, 147, 229, 230
Calderón, Alberto, 201
Cantor, Georg, 1, 5, 7, 8, 15, 31–33, 277
Cantor, Moritz, 20, 22, 24, 228
Carathéodory, Constantin, 93, 94, 101, 102, 104, 235
Carleman, Torsten, 93
Carleson, Lennart, xi, 118, 178, 199, 201, 204, 209–213, 215, 297
Carnal, Henri, 271, 272
Cartan, Élie, 79, 93, 94, 97, 101–103, 105, 235
Cartan, Henri, 128, 134, 140, 176, 177, 182, 185, 189, 190
Casacuberta, Carles, 292
Castelnuovo, Guido, 86, 127
Cayley, Arthur, 29, 47, 50
Chandrasekharan, Komaravolu, 56, 140, 147, 190, 194, 195
Chang, Sun-Yung Alice, 290
Chasles, Michel, 5, 19
Chebotaryov, N., 93
Chebyschev, Pafnuty, 18, 180
Cheeger, Jeff, 294
Chern, Shiing-Shen, 131, 145, 186, 287, 288, 291
Chevalley, Claude, 140, 145
Chorin, Alexandre J., 261
Christoffel, Erwin, 29
Church, Alonzo, 145, 177
Clebsch, Alfred, 5

Coates, John, 277
Cohen, Paul J., 117, 177
Coifman, Ronald, 297
Cole, Frank N., 7
congresses
    1893 Chicago, 6–8, 13
    1897 Zurich, 1, 9–18, 22, 23,
        25, 26, 28, 29, 35, 51, 55,
        56, 83, 92, 98, 106, 147,
        152, 155, 176, 227–228,
        231, 267, 269, 303
    1900 Paris, 15, 16, 19–26, 28,
        32, 35, 152, 156, 157, 189,
        190, 228, 231, 249
    1904 Heidelberg, 16, 24, 27–35,
        41, 43, 44, 51, 88, 120,
        152, 157, 190, 228, 231,
        239, 275
    1908 Rome, 34, 37–45, 48, 49,
        51–53, 77, 88, 120, 143,
        152, 158, 229, 232, 245,
        305, 306
    1912 Cambridge, 41, 45, 47–54,
        70, 72, 73, 77, 89, 90, 98,
        143, 152, 158, 229, 245,
        248, 305
    1916 Stockholm, 53, 70, 73,
        143
    1920 Strasbourg, 69–75, 77, 80,
        89, 90, 93, 106, 152, 159,
        229, 232, 305
    1924 New York, 73, 75, 76, 106
    1924 Toronto, 75–81, 89, 90,
        93, 96, 97, 106, 111, 112,
        116, 152, 159, 229, 238,
        305
    1928 Belgium, 73
    1928 Bologna, 83–92, 95, 97,
        145, 150, 152, 160, 229,
        232, 233, 239, 245, 305
    1932 Zurich, xii, xiii, 56, 89,
        91–99, 101, 102, 104–106,
        109, 112, 113, 129, 149,
        152, 161, 190, 219, 229,

        234, 245, 255, 263, 264,
        267, 272, 273, 305
    1936 Oslo, 56, 67, 94, 98,
        101–107, 109, 112, 113,
        115, 117, 118, 127, 129,
        130, 149, 151, 152, 161,
        219, 229, 234, 235, 254,
        305
    1940 Cambridge, MA, 106,
        125, 126, 131
    1950 Cambridge, MA, xv, 112,
        113, 117, 127–133, 135,
        149–153, 162, 223, 248
    1954 Amsterdam, xv, xvi, 60,
        61, 112, 113, 117,
        133–139, 141, 150–152,
        163, 229, 230, 234, 240,
        245, 246, 251, 252
    1958 Edinburgh, 81, 112, 113,
        116, 117, 138–143, 145,
        150–152, 163, 234, 246
    1962 Stockholm, xi, xii, 64, 112,
        113, 117, 118, 138,
        141–147, 150–152, 164,
        176, 234, 240, 246, 248,
        249
    1966 Moscow, xii, 56, 57, 64,
        112–114, 116, 117, 147,
        150, 164, 173, 175–183,
        189, 195, 196, 236, 241,
        242, 245, 259, 285
    1970 Nice, 113, 116, 117, 183,
        185–190, 195
    1974 Vancouver, 60, 61, 113,
        117, 165, 190–198, 202,
        203, 242, 243, 287
    1978 Helsinki, xi, 57, 113, 116,
        118, 165, 166, 198–205,
        236, 245–248
    1982 Warsaw, 63, 110, 117,
        119, 120, 167, 204, 205,
        207–216, 221, 223, 236,
        245, 248
    1986 Berkeley, 63, 113, 117,
        120, 168, 217, 219–225,

        237, 243–245, 249, 264,
        267
    1990 Kyoto, 61, 96, 97, 113,
        117, 120, 168, 225, 237,
        247, 248, 255, 257–267,
        271, 272
    1994 Zurich, xiii, 56, 63, 64,
        98, 114, 117, 120, 169,
        244, 246, 248, 255,
        266–273, 282
    1998 Berlin, 58, 114, 117, 120,
        170, 244, 245, 249, 250,
        255, 273, 275–284, 297
    2002 Beijing, 58, 59, 117, 120,
        171, 244, 245, 249, 255,
        284–292, 296
    2006 Madrid, xi, xiii, xiv, 59, 60,
        62, 63, 65, 116, 117, 120,
        123, 171, 181, 245,
        251–253, 255, 288,
        292–303
    2010 Hyderabad, 64, 65, 255,
        301, 302
    other proposals, 53, 98, 138,
        143, 204, 212, 267, 292
Connes, Alain, 110, 117, 201, 202,
    209, 211
Conway, John H., 270
Cook, Stephen A., 220, 261
Corput, Johannes G. van der, 104,
    133
Coxeter, Harold Scott MacDonald,
    191, 251
Crathorne, A. R., 68
Crelle, August Leopold, 4
Cremona, Luigi, 5, 29

d'Alembert, Jean, 1
Dantzig, Daniel van, 136
Darboux, Gaston, 25, 30, 39, 40
Darwin, Horace, 50
Darwin, Sir George H., 47, 48, 50,
    53, 229
Daubechies, Ingrid, 255, 271, 288
Davenport, Harold, 129, 140, 177

Dawson, Donald, 294
Debreu, Gérard, 196
Deift, Percy, 297
Deligne, Pierre, 64, 117, 196, 199, 201, 202, 220
Demailly, Jean-Pierre, 297
Deninger, Christopher, 279
Deuring, M., 177
DeVore, Ronald A., 297
Diaconis, Persi, 279
Dickson, Leonard Eugene, 70, 75, 79
Dieudonné, Jean, 135, 177, 183, 189, 190, 195
Dobrushin, R., 201, 202
Donaldson, Simon K., 117, 221, 222
Donoho, David L., 290
Doob, Joseph Leo, 129, 185
Douady, Adrien, 270
Douglas, Jesse, 101, 104, 117
Drinfeld, Vladimir G., 117, 260, 261
Duff, George, 196
Duistermaat, Johannes J., 277
Dunnington, G. Waldo, 56, 102, 234
Durán, Antonio J., 253, 300
Dyck, Walther von, 39, 40
Dynkin, Eugene Borisovich, 145, 146, 190

Eckmann, Beno, 145, 269, 270, 272
Edwards, Robert D., 201
Efimov, N. V., 178
Eichler, M., 199
Eilenberg, Samuel, 129, 140
Eisenhart, Luther P., 67, 75, 106
Eliashberg, Yakov, 297
Engel, Friedrich, 104
Engquist, Björn, 278, 298
Enriques, Federigo, 49
Erdős, Paul, 211, 284
Euler, Leonhard, 1, 13, 17, 18, 34, 43, 51, 55, 56
excursions, 231–238
    Cambridge Scientific Instrument Company, 49, 50
    Cow Palace, 237

Hadrian's Villa, 232
Linge battlefield, 232
Neckar River, 231
Niagara Falls, 78, 79
Oslo fjord, 234, 235
Rapperswyl, 227, 234
Ravenna, 232, 233
Transcontinental, 79, 116, 238
"Venetian Night", 228
exhibitions
    Ancient Chinese mathematical toys, 291
    Escher, 138, 251–252
    Exposition Universelle, 21–23, 26, 156, 157, 231
    Fractal Art: Beauty and Mathematics, 299
    History of Mathematical Knowledge, 298
    The ICM through History, xi, xiii, xiv, 302, 303
    The Life of Numbers, 251–254, 298
    Literature and Models, 32
    mathematical books in the Scottish National Library, 141
    Mathematik und Alltag, 281
    mathematical typography, 141
    rare books on mathematical sciences, 89
    Terror and Exile, 283

Föllmer, Hans, 296
Faddeev, Ludwig, 220, 225, 259–261, 290, 292
Faltings, Gerd, 117, 221, 222
Fefferman, Charles, 117, 196, 199, 201, 202, 260, 296
Fehr, Henri, 43, 51
Feigin, Boris L., 261
Feit, Walter, 186, 270
Fejér, Lipót, 98
Felder, Giovanni, 295
Feller, William, 140, 177

Fermat's Last Theorem, 114, 271, 273, 278
Fibonacci, Leonardo Pisano, 38, 59, 60, 89
Fields Medal, xi, xii, 97, 101, 109–119, 127, 194, 219, 303
    awardees, 116–117
    1936, 101, 104
    1950, 128, 134
    1954, 134
    1958, 140
    1962, 118, 145
    1966, 177
    1970, 185, 186
    1974, 194, 195
    1978, 199, 200, 202
    1982, 202, 209, 210
    1986, 221
    1990, 260, 265
    1994, 269, 270
    1998, 277
    2002, 287, 289
    2006, 293
    Committee, 97, 101, 102, 106, 110, 112, 113, 128, 134, 140, 143, 145, 149, 177, 185, 194, 199, 210, 220, 260, 269, 277, 287, 294
    Manilius, poet, 111
    McKenzie, Robert Tait, sculptor, 110, 112
    memorandum, 97, 109, 110, 112–114
    number of, 97, 110, 112, 113, 177, 194, 209, 221, 287
    problems, 116, 177, 181, 185, 186, 190, 195, 199, 200, 203, 204, 283, 293, 294
    under-forty rule, 110, 113–114, 177, 278
Fields, John Charles, 76, 77, 80, 97, 101, 109, 111, 112, 115–116

finances of ICM, 18, 26, 35, 38, 44, 73, 75, 88, 91, 106, 120, 125, 131, 138, 141, 143, 147, 190, 197, 202, 216, 224, 225, 265, 272, 292
Fleming, Wendell H., 211
Floer, Andreas, 261
Forsyth, Andrew R., 39, 45
Fröhlich, Jürg, 222, 271, 277
Fréchet, Maurice, 80, 86, 104, 128, 235
Fredholm, Ivar, 15
Freedman, Michael H., 117, 221, 277
Freudenthal, Hans, 189
Friedlander, Eric, 222
Friedrichs, Kurt O., 140
Frostman, Otto, 143, 145, 195
Fuchs, Lazarus, 29, 115
Fueter, Rudolf, 91–94, 102, 103

Gödel, Kurt, 1, 129
Galilei, Galileo, 41, 89
Galitzin, Prince B., 49
Gallavotti, Giovanni, 279
Gauss Prize, 120–123, 255, 288, 295, 303
    Arnold, Jan, artist, 121
    Ceres planetoid, 120–122
    Committee, 295
    winner, 295, 296
Gauss, Carl Friedrich, 56, 110, 120–123
Gehring, Frederick W., 222
Geiser, Karl F., 13, 15, 16, 18, 92
Gelfand, Israil Moiseevich, 136, 145, 146, 150, 186, 190
Gelfond, Alexandr, 106
Gergonne, Joseph Diaz, 4
Ghys, Étienne, 297
Gleason, Andrew, 219, 220
Glimm, James, 196, 220
Goddard, Peter, 277
Goldstein, S., 136
Goldwasser, Shafi, 288, 289
Golomb, Michael, 283

Gorenstein, Daniel, 201
Gowers, William Timothy, 117, 277, 278
Grötschel, Martin, 277, 284, 295
Graham, Ronald, 278
Grauert, Hans, 146
Graustein, W. C., 125
Greenhill, Alfred, 30, 32, 34, 43, 51
Greuel, Gert-Martin, 295
Griffiths, Phillip A., 187, 288
Gromov, Mikhail, 190, 222
Grothendieck, Alexander, 116, 117, 140, 177, 181, 185, 189
Guccia, Giovanni, 5, 41, 42
Guldberg, Alf, 98
Gårding, Lars, xi, 116, 140, 145

Hörmander, Lars, xi, 116–118, 145, 185, 187, 220
Haagerup, Uffe, 289
Hackbusch, Wolfgang, 279, 288
Hadamard, Jacques, 14, 43, 52, 86, 94, 105, 127, 149, 233, 234
Haller, Albrecht von, 99
Halsted, George Bruce, 15, 21, 23
Hamilton, Richard, 294, 297
Hamilton, Sir William Rowan, 80
Hardy, Godfrey H., 69, 72, 73, 81, 94, 106, 128, 234
Harish-Chandra, 135, 178
Hasse, Helmut, 104
Hausdorff, Felix, 15, 32
Hawking, Stephen, 291
Hecke, Erich, 103, 106, 127
Heilbronn, H. A., 190
Henrici, Peter, 146
Hermite, Charles, 7, 8, 18, 22, 29, 227
Hilbert, David, 7, 20, 21, 31, 32, 40, 83–86, 92, 106, 282
    twenty-three problems, 19–21, 32, 177, 189
Hille, Einar, 125

Hironaka, Heisuke, 116, 117, 185, 189, 260, 265
Hirzebruch, Friedrich, 140, 185, 204, 275, 277, 278, 282
history and culture
    Bjørnson, Bjørnstjerne, poet, 102
    Codex Vigilanus, 63, 253
    Cold War, 116, 149, 173, 196, 264
    Collet, Camilia, feminist, 103
    Darwin, Charles, 48
    Economic Depression of 1929, 91, 229
    Enzensberger, Hans Magnus, writer, 281, 282
    Escher, Maurits Cornelis, artist, 251
    French Revolution, 2
    Great War, 67–69, 72, 83, 91, 126, 143, 267, 305
        Allied Powers, 69, 73, 305
        Central Powers, 69, 72, 75, 83, 84, 305
    Haida people of western Canada, 60, 61
    Iron Curtain, 149, 151, 176
    martial law in Poland, 212, 213
    Marx, Karl, 7, 287
    Munch, Edvard, painter, 101
    Munch, Peter Andreas, historian, 102
    Nasar, Silvia, journalist, 294
    Nazi regime, 91, 282, 283
    Sartre, Jean-Paul, thinker, 110
    Scientific Revolution, 2
    Solidarność, 212, 213
    Treaty of Locarno, 83
    Treaty of Versailles, 70, 72
    Ushio, Keizo, sculptor, 299
    Vaa, Dyre, sculptor, 104
    Vietnam War, 181
    Vigeland, Gustav, sculptor, 103

World War II, xi, xv, 107, 125–127, 150, 207, 229, 265, 282, 305

World's Columbian Exposition, 1893, 6, 7

Hobson, Ernest W., 15

Hodge, William V. D., 128, 129, 138, 139, 142, 152

Hofer, Helmut W., 279, 287

Hollander, Frank den, 295

Hooley, Christopher, 211

Hopcroft, John, 296

Hopf, Heinz, 56, 113, 116, 128, 134, 140

Hopkins, Michael J., 290

Hopper, Grace, 131

Hrushovski, Ehud, 280

Hsiang, Wu-Chung, 211

Huisken, Gerhard, 294

Humboldt, Wilhelm von, 2

Hurewicz, Witold, 131

Hurwitz, Adolf, 1, 13, 16, 20, 106, 227

Håstad, Johan, 295

ICM miscellany
American way of life, 151
boat accommodation, 203
Borromean rings, 65, 296
Gaussian primes, 137, 138
"hippie ICM", 242
Keizo Ushio's sculpture, 299
Kusudama, 61
Mathematical Tripos, 48
Maya language, 146, 147
Pascal's performance, 229
soccer game, 183
Wreck Beach, 198

Ihara, Yasutaka, 261, 287

Ingraham, M. H., 125

institutions
Air Force Office of Scientific Research, 224
Army Research Office, 224
Carnegie Corporation, 131

International Association for the Promotion of Cooperation with Scientists from the Independent States of the Former Soviet Union, INTAS, 272

International Council of Scientific Unions, ICSU, 305

International Research Council, 69, 70, 72, 73, 75, 76, 81, 83, 84, 97, 229, 304, 305

International Union of History and Philosophy of Science, 306

National Research Council of Canada, 197

National Science Foundation, 198, 224

Office of Naval Research, 224

Rockefeller Foundation, 131

Royal Canadian Mint, 110, 112, 303

Science Council of Japan, 265

Sir Dorabji Tata Trust, 113

Soros Foundation, 272

Vaughn Foundation Fund, 224

International Commission on Mathematical Instruction, 43, 88, 94, 105, 136, 146, 189, 223, 270, 306

Central Committee, 51

International Congress on Mathematical Education, 189, 306

International Commission on the History of Mathematics, 270, 297

International Congress of Mathematicians, xi, 1, 9, 27, 34, 55, 64, 74, 80, 110, 125, 149, 205, 305
acronym, 58, 64
group photographs, 81, 96, 102, 103, 134, 135, 152, 153

International Congress of Mathematics, 74

International Mathematical Congress, 75, 77, 80
numbering, 77, 89, 93
regulations, 13, 16, 31, 43, 227
Soviet participation, 77, 95, 106, 149–150, 177, 190, 196, 203, 211, 212, 214, 224, 225
tribute to the past of mathematics, 17, 27–29, 38, 50, 80, 87, 104, 138, 207, 219, 282

International Mathematical Congresses: An Illustrated History 1893–1986, xi, 223

International Mathematical Union, xv, 110, 111, 113, 114, 116, 120, 134, 138, 141, 143, 147, 150, 177, 182, 188, 195, 196, 210–213, 216, 224, 269, 278, 288, 301, 305–306
archives, xi, 113, 267
China's membership, 224
Digital Mathematical Library, 301
International Association of Mathematicians, 43, 51, 305
logo, 65, 296
presidents, xiii, 56, 113, 118, 134, 140, 145, 147, 182, 190, 198, 204, 220, 225, 259, 266, 273, 284, 292, 301
re-foundation of, 125, 126, 131, 147, 149, 305
statutes, 131, 224
World Directory of Mathematicians, 17, 147, 306

World Mathematical Year 2000, 284, 306

Itô, Junko, 296
Itô, Kiyoshi, 123, 220, 259, 265, 295, 296
Iwaniec, Henryk, 297
Iwasawa, Kenkichi, 260
Iyanaga, S., 185

Jablonskii, S. V., 196
Jacobi, Carl Gustav Jacob, 27–29, 35
James, I. M., 199, 201
Jessen, Børge, 94, 136
Jimbo, Michio, 260
Johnstone, Iain M., 297
Jones, Vaughan F. R., 117, 260, 261
Jordan, Camille, 70, 74
journals
    Acta Eruditorum, 3
    Acta Mathematica, 53
    The American Mathematical Monthly, 15, 21, 54
    Annales de Mathématiques Pures et Appliquées, 4
    Annali di Matematica Pura ed Applicata, 87
    Bulletin de la Société Mathématique de France, 5
    Bulletin of the American Mathematical Society, 21, 35, 125, 231, 232, 245
    Commentarii Mathematici Helvetici, 95
    Deutsche Mathematik, 91
    Fundamenta Mathematicae, 209, 217
    Giornale de' letterati, 3
    Jahrbuch über die Fortschritte der Mathematik, 5, 16
    Jahresberichte der Deutsche Mathematische -Vereinigung, 84
    Journal de Mathématiques Pures et Appliquées, 4
    Journal des Savants, 2
    Journal für die reine und angewandte Mathematik (Crelle's Journal), 4
    L'Enseignement Mathématique, 43
    L'Intermédiaire des Mathématiciens, 8
    Matematicheskii Sbornik, 5
    Mathematische Annalen, 5
    Nature, 22, 23
    Notices of the American Mathematical Society, 278
    Philosophical Transactions of the Royal Society, 3
    Proceedings of the London Mathematical Society, 5
    Rendiconti del Circolo Matematico di Palermo, 5, 42
    Repertoire Bibliographique des Sciences Mathématiques, 5, 16
    Revue Semestrielle des Publications Mathématiques, 5
    Science, 7, 15, 68, 95
    Studia Mathematicae, 217
    Transactions of the American Mathematical Society, 5
Julia, Gaston, 67, 93, 98, 102, 106, 107, 138, 239

König, Jules, 31, 32
Königsberger, Leo, 29
Kármán, Theodore von, 86
Kac, Victor, 289
Kahane, Jean-Pierre, 146
Kakutani, Shizuo, 130
Kashiwara, Masaki, 201, 269
Kato, Kazuya, 297
Kato, Tosio, 187
Katz, N., 201
Keisler, Jerome, 187
Keldysh, Mstislav Vsevolodovich, 175, 177
Keller, Joseph B., 271
Kesten, Harry, 290

Khintchine, Alexandr Yakovlevich, 106
Kirwan, Frances, 290
Kleene, Stephen C., 129, 140
Klein, Felix, 5–8, 14–16, 32, 34, 40, 43, 51, 88, 128
Kleinberg, Jon, 120, 295, 296
Kodaira, Kunihiko, 116, 117, 134, 194, 264, 284
Koenigs, Gabriel, 72, 73, 84
Kohn, Robert V., 297
Kolmogorov, Andrey Nikolaevich, 128, 135, 140, 146, 149, 150, 181
Komatsu, Hikosaburo, 257, 264, 265
Kontsevich, Maxim, 117, 271, 277, 278
Kosambi, D., 128
Kowalewski, G., 32
Krasovich, N. N., 201, 202
Krein, Mark Grigorievich, 178
Kreiss, Heinz-Otto, 196
Kuratowski, Kazimierz, 209

Lafforgue, Laurent, 117, 287–290
Lagrange, Joseph Louis, 1, 74, 121
Laisant, Charles, 8
Lalande, Jérôme, astronomer, 123
Lanczos, Cornelius, 140
Landau, Edmund, 49, 282
Langlands, Robert P., 201
language issue at ICM, 12, 16, 23, 29, 51, 52, 67, 87, 88, 126, 134, 189, 290
Laplace, Pierre Simon, 120
Larmor, Sir Joseph, 49, 70
Laumon, Gérard, 288
Lavrentyev, Mikhail Alekseevich, 147, 175, 177
Lawson, H. Blaine, 287
Lax, Peter D., 211, 260
León, Manuel de, 296
Lebowitz, Joel, 282
Ledermann, Walter, 283
Lefschetz, Solomon, 104, 106

Lehto, Olli, xi, xiii, 195, 199, 201,
    205, 278, 283, 284, 305
Leibniz, Gottfried Wilhelm, 3, 33,
    55, 202
    calculating machine, 33
Leighton, Tom, 278
Lemoine, Émile, 8
Lenstra, Hendrik W., 222, 269
Leray, Jean, 185
Levi-Civita, Tullio, 15, 106, 127
Le Roux, Jean, 80
Lichnerowicz, André, 136, 189
Lie, Sophus, 29, 67, 104, 105
Lindelöf, Ernst, 15, 25, 104, 118
Line, J. R., 125
Lions, Jacques-Louis, 196, 210, 266,
    269
Lions, Pierre-Louis, 117, 269–271
Liouville, Joseph, 4
Littlewood, John E., 94
Lobachevsky, Nikolai, 180
logotypes
    ICM, 56–61, 63–65
    IMU, 65, 296
Lord Rayleigh, John W. Strutt, 48
Lorentz, Hendrik Antoon, 38, 39
Lott, John, 295
Lovász, László, 261, 263, 301
Lusin, Nikolai N., 86, 178
Lusztig, George, 263

Macdonald, I. G., 280
Mackey, George W., 146
MacPherson, Robert D., 211, 277
Madsen, Ib, 297
Majda, Andrew J., 263
Malcev, A. I., 178
Malgrange, Bernard, 178, 194
Mallat, Stéphane, 280, 295
Malliavin, Paul, 210
Mandelbrot, Benoit, 299
Manin, Yuri I., 190, 201, 202, 260,
    277, 278, 297
Marchuk, G. I., 187
Marcolongo, Roberto, 86

Margulis, Gregori, 116, 117,
    199–201, 203, 204, 263,
    283
Markov, Andrey, 18
Maschke, Heinrich, 7
Maslov, V. P., 211
mathematical encyclopedias
    Encyclopédie des sciences
        mathématiques, 34
    Encyklopädie der Mathematischen
        Wissenschaften, 5, 34
mathematical societies, 4
    American Mathematical Society,
        xv, 5, 7, 26, 28, 75, 81,
        112, 125, 127, 131, 198,
        224, 225
    Bosnian Mathematical Society,
        272
    Canadian Mathematical
        Congress, 197
    Chinese Mathematical Society,
        284
    Circolo Matematico di Palermo,
        5, 28, 34, 41, 42, 112
    Deutsche
        Mathematiker-Vereinigung,
        5, 8, 16, 23, 27, 28, 35, 51,
        84, 93, 112, 120, 126, 239,
        267, 276, 282, 283, 288,
        303
    Edinburgh Mathematical
        Society, 5
    European Mathematical Society,
        298
    Kunstrechnungs Liebende
        Gesellschaft, 5
    London Mathematical Society,
        5, 28, 45, 97, 109
    Mathematical Association of
        America, 131, 225
    Mathematical Society of Japan,
        265
    Moscow Mathematical Society, 5
    New York Mathematical Society,
        5

Polish Mathematical Society, 212
Real Sociedad Matemática
    Española, 300, 301
Société Mathématique de
    France, 5, 15, 28, 112
Société Mathématique du
    Canada, 190
Société Mathématique Suisse,
    112
Societat Catalana de
    Matemàtiques, 301
Society for Industrial and
    Applied Mathematics, 225
St. Petersburg Mathematical
    Society, 283
Unione Matematica Italiana, 229
Wiskundig Genootschap, 5, 28,
    133, 136
Mathematics without Borders: A
    History of the International
    Mathematical Union, xi,
    283, 284, 305
Matiyasevich, Yuri V., 189, 269, 270
Mazur, Barry, 211, 221, 269
McDuff, Dusa, 280
McMullen, Curtis T., 117, 277, 278,
    294
Mehrtens, Herbert, 282
Mellin, Hjalmar, 15
Melrose, Richard B., 263
Menger, Karl, 93
Mergelyan, Sergey, 175
Mertens, Franz, 15
Milner, Eric C., 196
Milnor, John, xi, 117, 118, 145, 146,
    185, 220, 221
Minkowski, Hermann, 7, 15, 20, 33
Mittag-Leffler, Gösta, 15, 18, 24, 25,
    40, 53, 69, 72, 73, 116,
    143, 227
Miwa, Tetsuji, 280
Modern Mathematics in the Light of the
    Fields Medals, 118
Monastyrsky, Michael, 118
Montel, Paul, 185

Montgomery, Deane, 198, 199
Moore, C. L. E., 232
Moore, Eliakim Hastings, 7
Moore, M., 34
Mordell, Louis Joel, 98, 104
Morgan, John W., 294
Mori, Shigefumi, 116, 117, 260, 263
Morse, Marston, 93, 125, 128–130
Moser, Jürgen, 56, 199, 220, 225, 280
Mostowski, Andrzej, 194
Motchane, Léon, 177
Mumford, David, 117, 194, 195, 210, 269, 273, 278, 289
music at ICM, 92, 134, 139, 143, 199–201, 209, 210, 219–221, 227, 234, 239, 242, 245–250, 258–260, 266, 268–270, 277, 279, 287, 296

Nörlund, Niels Erik, 70, 74
Nagata, Masayoshi, 225
Nakajima, Hiraku, 290
Nash, John F. Jr., 291, 294
Nemirovski, Arkadi, 297
Neugebauer, Otto, 104
Neumann, Bernhard, 283
Neumann, John von, 130, 135
Neunzert, Helmut, 297
Nevanlinna Prize, 118–119, 210, 303
    Committee, 210, 220, 261, 269, 278, 288, 295
    Heino, Raimo, sculptor, 119
    winners, 119, 210, 220, 261, 265, 269, 278, 288, 289, 293, 295
Nevanlinna, Rolf, 93, 104, 118–119, 145, 147, 198–200, 205
Newcomb, Simon, 39
Newman, Charles M., 296
Newman, M. H. A., 146
Newton, Isaac, 47, 55, 99, 110
Neyman, Jerzy, 136
Nielsen, Jakob, 103

Nikolsky, Sergey Mikhailovich, 136, 150
Nirenberg, Louis, 146, 210, 211
Noether, Emmy, 44, 88, 93, 96, 97, 255, 282
Noether, Fritz, 106
Noether, Max, 7, 15, 41
Norwood, G., 111
Novikov, Sergei, 116, 117, 185, 186, 190, 202, 220, 283, 287
numerals
    Chinese, 59
    Hindu-Arabic, 62, 63, 253, 254
    Roman, 58

Okounkov, Andrei, 117, 293, 295
Olech, Czewsław, 209, 211
Ore, Øystein, 104
Orlicz, Władysław, 209
Oseen, C. W., 104
Ostrowski, A., 134

Pólya, George, 129, 130, 233, 234, 239, 251
Padé, Henri, 15
Painlevé, Paul, 30
Palis, Jacob, 284, 288
Papanicolaou, George, 280, 287
Parshin, Aleksei N., 294
participation in ICM, 15, 25, 28, 35, 43, 44, 52, 53, 72, 77, 78, 85, 88, 95, 97, 105, 149–152, 173, 175, 176, 190, 197, 198, 202, 213, 223, 224, 264, 273, 283, 285, 292, 301
    countries, 15, 25, 35, 44, 53, 72, 77, 85, 97, 105, 151, 176, 197, 203, 214, 224, 264, 272, 283, 301
    publishers, 18, 26, 28, 29, 33–35, 44, 52, 87, 147
    women, 7, 16, 25, 35, 43, 52, 88, 96, 97, 105, 150, 255, 263, 271, 283

Pauli, Wolfgang, 94
Peano, Giuseppe, 15
Penrose, Roger, 202, 251
Perelman, Grigory, 116, 117, 293, 294, 296
Petrovsky, Ivan Georgievich, 175, 176
Pełczyński, Aleksander, 211
Piatetski-Shapiro, I. I., 178
Piazzi, Giuseppe, astronomer, 121
Picard, Émile, 15, 39, 40, 69, 70, 73, 74, 76, 83, 84, 92, 106
Piene, Ragni, 292
Pierpoint, James, 80
Pincherle, Salvatore, 7, 15, 80, 81, 83, 84, 89, 145
Pisier, Gilles, 280
Plancherel, Michel, 98
Pleijel, Å., 134
Poincaré, Henri, 7, 8, 13, 14, 22–25, 40, 41, 50, 138
Pol, Balthasar van der, 137, 138
Pontryagin, Lev Semenovich, 140, 187, 189, 194, 196, 203, 204
Poovey, Mary, 291
Popa, Sorin, 297
popularization of mathematics
    lectures at Urania, 281
    lectures in Beijing, 291
    panel discussion in Madrid, 298
posters of ICMs, xi, 55, 58, 59, 61, 62, 64, 196, 216, 217, 281
Postnikov, A. G., 178
Pringsheim, Alfred, 7, 15, 282
proceedings of ICM, xi, xiii, xv, xvi, 14, 16, 19, 27, 32, 41–43, 55, 56, 64, 74, 76, 77, 83, 85, 89, 94, 97, 101, 104, 107, 111, 131, 140, 141, 152, 175, 179, 207, 227, 275, 297
Prohorov, J. V., 199
Puppini, Umberto, 86

Quarteroni, Alfio, 297
Quillen, Daniel G., 117, 196, 199, 201, 202

Rabin, Michael, 211, 212, 261, 288
Ramanujan, Srinivasa, 64
Ratner, Marina, 255, 271
Razborov, Alexander A., 120, 261, 278
Reid, Constance, xi, 85, 223
Rham, Georges de, 113, 114, 147, 176, 177, 182
Richardson, R. G. D., 125
Riesz, Frédéric, 93
Ritt, Joseph F., 129
Rome, Adolfe, 129
Roth, Klaus Friedrich, 117, 140
Rudin, Mary Ellen, 219, 263
Rudio, Ferdinand, 16–18, 51, 147, 227
Ruelle, David, 211
Runge, Carl, 33

Saito, Kyoji, 277
Salmon, George, 29
Salomaa, A., 210
Sarnak, Peter, 280, 297
satellite conferences, 291, 297
Sato, Mikio, 211
Sautoy, Marcus du, 298
Schönflies, Arthur, 7, 15, 31, 32
Schönhage, Arnold, 222
Schütte, Kurt, 178
Schiffer, Menahem, 140
Schinzel, Andrzej, 210
Schmid, Wilfried, 202
Schmidt, Erhard, 67
Schmidt, Wolfgang M., 196
Schoen, Richard, 222
Schouten, Jan A., 133, 134, 152
Schröder, Johann, 178
Schramm, Oded, 297
Schrijver, Alexander, 269, 288
Schwartz, Jacob, 210

Schwartz, Laurent, 117, 128, 129, 140, 149, 189
Schwarz, Hans Amandus, 27, 29, 115
scientific institutes
    Institut des Hautes Études Scientifiques, 177
    Institute for Advanced Study, 131
    Isaac Newton Institute, 114
    Konrad-Zuse-Center, 277
    Mathematical Sciences Research Institute, 220
    Mathematischesforschungsinstitut Oberwolfach, 151
    Research Institute for Mathematical Sciences, 264
    Stefan Banach International Mathematical Center, 209
    Steklov Institute, 294
    Tata Institute of Fundamental Research, 147
    Weierstrass Institute for Applied Analysis and Stochastics, 277
scientific structure of ICM
    communications, 136, 140, 145, 152, 179, 180, 187, 190, 195, 202, 212, 222, 223, 263, 264, 279, 289, 296
    plenary lectures, xvi, 13, 15, 16, 20, 22, 29, 38, 39, 43, 48, 49, 70, 71, 79, 85, 86, 89, 91, 93, 94, 96, 97, 103, 105, 106, 128, 129, 135, 140, 145, 146, 150–152, 178, 186, 189, 190, 195, 196, 201–203, 211, 212, 222, 232, 255, 261, 263, 270, 271, 273, 279, 281, 289, 290, 294, 296, 297, 300
    section lectures, 13, 15, 16, 31, 43, 71, 80, 87, 94, 103, 106, 129, 130, 136, 140,

145, 152, 179, 180, 187, 190, 195, 197, 202, 203, 211, 222, 263, 270, 279, 289, 296
    sections, xvi, 13, 14, 16, 23, 30, 31, 40, 49, 71, 77, 87, 94, 104, 130, 136, 141, 145, 179, 187–189, 195, 197, 202, 212, 222, 263, 279, 290, 296
Scott, Charlotte Angas, 16, 21, 23, 24, 26
Segre, Beniamino, 129, 136
Segre, Corrado, 15, 30, 41
Seidel, J. J., 251
Selberg, Atle, 117, 128, 129, 146
Serre, Jean-Pierre, 117, 134, 146, 177, 189
Seshadri, C. S., 220
Severi, Francesco, 41, 42, 80, 93, 94, 97, 98, 101, 106
Seymour, Paul, 271
Shafarevich, Igor Rostislavovich, 146, 185, 190, 260
Shelah, Saharon, 211, 212, 222
Shiryaev, A. N., 202
Shor, Peter W., 120, 278, 280
Siegel, Carl Ludwig, 103, 140
Sierpiński, Wacław, 91, 93, 104
Sigmund, Karl, 280
Sinai, Yakov G., 190, 263, 287
Singer, I.M., 196
Siu, Yum-Tong, 212, 290
Skorohod, A. V., 222
Sloan, Ian, 295
Smale, Stephen, 117, 177, 178, 181–182, 222, 277, 278
Smith, David Eugene, 43, 51, 88, 95, 111
Snyder, Virgil, 245
Sobolev, Sergey Lvovich, 146, 147
Sossinsky, Alexei B., 298
Soulé, Christophe, 288
Spencer, Tom, 294

stamps, ICM commemorative, 57–59, 61–63, 273, 300
Stanley, Richard P., 297
Steenrod, Norman E., 140
Stein, Charles M., 178
Stein, Elias M., 187, 222
Steiner, Jakob, 13, 56, 109
Steklov, W., 77
Stenzel, J., 93, 94
Stiefel, Edward Ludwig, 135
Stone, Marshall H., 125, 126, 131, 225
Strassen, Volker, 222, 261
Study, Edward, 7
Størmer, Carl, 39, 80, 102, 103, 106, 127
Sudan, Madhu, 120, 288
Sullivan, Dennis, 196, 269
Sullivan, John, 65, 296
Suslin, Andrei A., 222
Swan, Richard G., 187
Sylvester, James, 29, 47
Synge, John Lighton, 97, 111, 112
Szökefalvi-Nagy, B., 199
Szegö, Gábor, 134

Takagi, Teiji, 97, 101
Talagrand, Michel, 280
Tandon, Rajat, 301
Tannery, Jules, 22
Tannery, Paul, 51
Tao, Terence, 116, 117, 293, 296, 297
Tarjan, Robert, 119, 120, 210, 211, 269
Tarski, Alfred, 129, 136
Tate, John, 187, 194, 195
Tauber, Alfred, 15
Taubes, Clifford H., 271, 278
Taylor, Richard, 290
Temple, George, 140
Terradas, Esteban, 98
Theilheimer, Feodor, 283
Thom, René, 117, 140, 177, 212

Thompson, John G., 117, 178, 185, 186
Thurston, William P., 117, 202, 209, 210
Tian, Gang, 289
Tikhonov, Andrei N., 181
Tisseyre, François, 298
Titchmarsh, Edward C., 134, 135
Tits, Jacques, 146, 196, 199, 201, 203, 204, 269
Tondeur, Philippe, 298
Tonelli, Leonida, 86
Turán, P., 185
Tyler, H. W., 35, 231, 239

Uhlenbeck, Georeg E., 141
Uhlenbeck, Karen, 255, 263
Union Mathématique Internationale, 70, 71, 73, 75, 76, 80, 81, 83, 84, 88, 89, 97, 98, 106, 304–306
    dissolution, 98, 110
    foundation, 73
    presidents, 71, 75, 80, 81, 83, 84, 89, 97
    statutes, 70, 97, 304
Urbanik, Kazimierz, 205

Vázquez, Juan Luis, 297
Vafa, Cumrun, 280
Valiant, Leslie, 120, 220, 222
Valiron, Georges, 93
Vallée Poussin, Charles de la, 14, 70, 71, 75, 76, 80, 98, 127, 239
Varadhan, S.R. Srinivasa, 269–271
Varchenko, Alexandre, 263
Vassilief, A., 8, 23, 34
Vassiliev, Victor A., 271
Veblen, Oswald, xv, 86, 91, 98, 103, 127, 133, 152
Vergne, Michèle, 294, 297
Veronese, Giuseppe, 40
Vershik, Anatoly M., 283

Viana, Marcelo, 280
Vinogradov, Ivan Matveevich, 175, 178, 190
Vitushkin, A. G., 196
Vladimirov, V. S., 140
Voevodsky, Vladimir, 117, 280, 288
Vogan, David A., 222
Voiculescu, Dan, 271
Volterra, Vito, 15, 23, 25, 34, 40, 69–71, 86, 106

Wald, Abraham, 129
Wall, C. T. C., 187, 210, 211
Wavre, Rolin, 93
Weber, Heinrich, 27–29, 34
Weierstrass, Karl, 24, 29, 115
Weil, André, 130, 136, 202, 284
Werner, Wendelin, 116, 117, 294–296
Weyl, Hermann, 86, 91, 93, 98, 99, 134, 235, 282
White, Henry S., 7
White, Sir W. H., 49
Whitney, Hassler, 129, 131, 145
Whittaker, Sir Edmund, 139
Wielandt, Helmut, 141
Wiener, Norbert, 101, 102, 104, 130, 235
Wigderson, Avi, 120, 269, 270, 297
Wilder, Raymond L., 128
Wiles, Andrew, 114, 271, 273, 278, 279
Winograd, S., 220
Wirtinger, Wilhelm, 30, 31
Witten, Edward, 117, 222, 260, 261, 290
Wright, Margaret H., 295
Wu, Wen-Tsun , 287, 291

Yamaguti, M., 269
Yang, Paul C., 290
Yau, Shing-Tung, 117, 202, 209, 211

Yoccoz, Jean-Chistophe, 117, 269–271
Yosida, Kosaku, 136, 145
Young, J. W. A., 54
Young, William Henry, 80, 86, 97

Zack, Franz, astronomer, 122, 123
Zaremba, Stanislaw, 98
Zariski, Oscar, 130, 140
Zelmanov, Efim, 117, 269, 270, 287
Zermelo, Ernst, 32

Zhao, Shuang, proof of Pythagorean theorem, 58
Zhiming, Ma, 292
Zizcenko, A. B., 204
Zygmund, Antoni, 194

T - #0494 - 071024 - C344 - 229/204/16 - PB - 9780367385965 - Gloss Lamination